COTTAGE ECONOMY

WILLIAM COBBETT

INTRODUCTION.

TO THE LABOURING CLASSES OF THIS KINGDOM.

1. Throughout this little work, I shall number the Paragraphs, in order to be able, at some stages of the work, to refer, with the more facility, to parts that have gone before. The last Number will contain an Index, by the means of which the several matters may be turned to without loss of time; for, when economy is the subject, time is a thing which ought by no means to be overlooked.

2. The word Economy, like a great many others, has, in its application, been very much abused. It is generally used as if it meant parsimony, stinginess, or niggardliness; and, at best, merely the refraining from expending money. Hence misers and close-fisted men disguise their propensity and conduct under the name of economy; whereas the most liberal disposition, a disposition precisely the contrary of that of the miser, is perfectly consistent with economy.

3. ECONOMY means management, and nothing more; and it is generally applied to the affairs of a house and family, which affairs are an object of the greatest importance, whether as relating to individuals or to a nation. A nation is made powerful and to be honoured in the world, not so much by the number of its people as by the ability and character of that people; and the ability and character of a people depend, in a great measure, upon the economy of the several families, which, all taken together, make up the nation. There never yet was, and never will be, a nation permanently great, consisting, for the greater part, of wretched and miserable families.

4. In every view of the matter, therefore, it is desirable; that the families of which a nation consists should be happily off: and as this depends, in a great degree, upon the management of their concerns, the present work is intended to convey, to the families of the labouring classes in particular, such information as I think may be useful with regard to that management.

5. I lay it down as a maxim, that for a family to be happy, they must be well supplied with food and raiment. It is a sorry effort that people make to persuade others, or to persuade themselves, that they can be happy in a state of want of the necessaries of life. The doctrines which fanaticism preaches, and which teach men to be content with poverty, have a very pernicious tendency, and are calculated to favour tyrants by giving them passive slaves. To live well, to enjoy all things that make life pleasant, is the right of every man who constantly uses his strength judiciously and lawfully. It is to blaspheme God to suppose, that he created man to be miserable, to hunger, thirst, and perish with cold, in the midst of that abundance which is the fruit of their own labour. Instead, therefore, of applauding "happy poverty," which applause is so much the fashion of the present day, I despise the man that is poor and contented; for, such

content is a certain proof of a base disposition, a disposition which is the enemy of all industry, all exertion, all love of independence.

6. Let it be understood, however, that, by poverty, I mean real want, a real insufficiency of the food and raiment and lodging necessary to health and decency; and not that imaginary poverty, of which some persons complain. The man who, by his own and his family's labour, can provide a sufficiency of food and raiment, and a comfortable dwelling-place, is not a poor man. There must be different ranks and degrees in every civil society, and, indeed, so it is even amongst the savage tribes. There must be different degrees of wealth; some must have more than others; and the richest must be a great deal richer than the least rich. But it is necessary to the very existence of a people, that nine out of ten should live wholly by the sweat of their brow; and, is it not degrading to human nature, that all the nine-tenths should be called poor; and, what is still worse, call themselves poor, and be contented in that degraded state?

7. The laws, the economy, or management, of a state may be such as to render it impossible for the labourer, however skilful and industrious, to maintain his family in health and decency; and such has, for many years past, been the management of the affairs of this once truly great and happy land. A system of paper-money, the effect of which was to take from the labourer the half of his earnings, was what no industry and care could make head against. I do not pretend that this system was adopted by design. But, no matter for the cause; such was the effect.

8. Better times, however, are approaching. The labourer now appears likely to obtain that hire of which he is worthy; and, therefore, this appears to me to be the time to press upon him the duty of using his best exertions for the rearing of his family in a manner that must give him the best security for happiness to himself, his wife and children, and to make him, in all respects, what his forefathers were. The people of England have been famed, in all ages, for their good living; for the abundance of their food and goodness of their attire. The old sayings about English roast beef and plum-pudding, and about English hospitality, had not their foundation in nothing. And, in spite of all refinements of sickly minds, it is abundant living amongst the people at large, which is the great test of good government, and the surest basis of national greatness and security.

9. If the labourer have his fair wages; if there be no false weights and measures, whether of money or of goods, by which he is defrauded; if the laws be equal in their effect upon all men: if he be called upon for no more than his due share of the expenses necessary to support the government and defend the country, he has no reason to complain. If the largeness of his family demand extraordinary labour and care, these are due from him to it. He is the cause of the existence of that family; and, therefore, he is not, except in cases of accidental calamity, to throw upon others the burden of supporting it. Besides, "little children are as arrows in the hands of the giant, and blessed is the man that hath his quiver full of them." That is to say, children, if they bring their cares, bring also their pleasures and solid advantages. They become, very soon, so many assistants and props to the parents, who, when old age comes on, are amply repaid for all the toils and all the cares that children have occasioned in their infancy. To be without sure and safe friends in the world makes life not worth having; and whom can we be so sure of as of our children? Brothers and sisters are a mutual support. We see them, in almost every case, grow up into prosperity, when they act the part that the impulses of nature prescribe. When cordially

united, a father and sons, or a family of brothers and sisters, may, in almost any state of life, set what is called misfortune at defiance.

10. These considerations are much more than enough to sweeten the toils and cares of parents, and to make them regard every additional child as an additional blessing. But, that children may be a blessing and not a curse, care must be taken of their education. This word has, of late years, been so perverted, so corrupted; so abused, in its application, that I am almost afraid to use it here. Yet I must not suffer it to be usurped by cant and tyranny. I must use it: but not without clearly saying what I mean.

11. Education means breeding up, bringing up, or rearing up; and nothing more. This includes every thing with regard to the mind as well as the body of a child; but, of late years, it has been so used as to have no sense applied to it but that of book-learning, with which, nine times out of ten, it has nothing at all to do. It is, indeed, proper, and it is the duty of all parents, to teach, or cause to be taught, their children as much as they can of books, after, and not before, all the measures are safely taken for enabling them to get their living by labour, or for providing them a living without labour, and that, too, out of the means obtained and secured by the parents out of their own income. The taste of the times is, unhappily, to give to children something of book-learning, with a view of placing them to live, in some way or other, upon the labour of other people. Very seldom, comparatively speaking, has this succeeded, even during the wasteful public expenditure of the last thirty years; and, in the times that are approaching, it cannot, I thank God, succeed at all. When the project has failed, what disappointment, mortification and misery, to both parent and child! The latter is spoiled as a labourer: his book-learning has only made him conceited: into some course of desperation he falls; and the end is but too often not only wretched but ignominious.

12. Understand me clearly here, however; for it is the duty of parents to give, if they be able, book-learning to their children, having first taken care to make them capable of earning their living by bodily labour. When that object has once been secured, the other may, if the ability remain, be attended to. But I am wholly against children wasting their time in the idleness of what is called education; and particularly in schools over which the parents have no control, and where nothing is taught but the rudiments of servility, pauperism and slavery.

13. The education that I have in view is, therefore, of a very different kind. You should bear constantly in mind, that nine-tenths of us are, from the very nature and necessities of the world, born to gain our livelihood by the sweat of our brow. What reason have we, then, to presume, that our children are not to do the same? If they be, as now and then one will be, endued with extraordinary powers of mind, those powers may have an opportunity of developing themselves; and if they never have that opportunity, the harm is not very great to us or to them. Nor does it hence follow that the descendants of labourers are always to be labourers. The path upwards is steep and long, to be sure. Industry, care, skill, excellence, in the present parent, lay the foundation of a rise, under more favourable circumstances, for his children. The children of these take another rise; and, by-and-by, the descendants of the present labourer become gentlemen.

14. This is the natural progress. It is by attempting to reach the top at a single leap that so much misery is produced in the world; and the propensity to make such attempts has been

cherished and encouraged by the strange projects that we have witnessed of late years for making the labourers virtuous and happy by giving them what is called education. The education which I speak of consists in bringing children up to labour with steadiness, with care, and with skill; to show them how to do as many useful things as possible; to teach them to do them all in the best manner; to set them an example in industry, sobriety, cleanliness, and neatness; to make all these habitual to them, so that they never shall be liable to fall into the contrary; to let them always see a good living proceeding from labour, and thus to remove from them the temptation to get at the goods of others by violent or fraudulent means; and to keep far from their minds all the inducements to hypocrisy and deceit.

15. And, bear in mind, that if the state of the labourer has its disadvantages when compared with other callings and conditions of life, it has also its advantages. It is free from the torments of ambition, and from a great part of the causes of ill-health, for which not all the riches in the world and all the circumstances of high rank are a compensation. The able and prudent labourer is always safe, at the least; and that is what few men are who are lifted above him. They have losses and crosses to fear, the very thought of which never enters his mind, if he act well his part towards himself, his family and his neighbour.

16. But, the basis of good to him, is steady and skilful labour. To assist him in the pursuit of this labour, and in the turning of it to the best account, are the principal objects of the present little work. I propose to treat of brewing Beer, making Bread, keeping Cows and Pigs, rearing Poultry, and of other matters; and to show, that, while, from a very small piece of ground a large part of the food of a considerable family may be raised, the very act of raising it will be the best possible foundation of education of the children of the labourer; that it will teach them a great number of useful things, add greatly to their value when they go forth from their father's home, make them start in life with all possible advantages, and give them the best chance of leading happy lives. And is it not much more rational for parents to be employed in teaching their children how to cultivate a garden, to feed and rear animals, to make bread, beer, bacon, butter and cheese, and to be able to do these things for themselves, or for others, than to leave them to prowl about the lanes and commons, or to mope at the heels of some crafty, sleekheaded pretended saint, who while he extracts the last penny from their pockets, bids them be contented with their misery, and promises them, in exchange for their pence, everlasting glory in the world to come? It is upon the hungry and the wretched that the fanatic works. The dejected and forlorn are his prey. As an ailing carcass engenders vermin, a pauperized community engenders teachers of fanaticism, the very foundation of whose doctrines is, that we are to care nothing about this world, and that all our labours and exertions are in vain.

17. The man, who is doing well, who is in good health, who has a blooming and dutiful and cheerful and happy family about him, and who passes his day of rest amongst them, is not to be made to believe, that he was born to be miserable, and that poverty, the natural and just reward of laziness, is to secure him a crown of glory. Far be it from me to recommend a disregard of even outward observances as to matters of religion; but, can it be religion to believe that God hath made us to be wretched and dejected? Can it be religion to regard, as marks of his grace, the poverty and misery that almost invariably attend our neglect to use the means of obtaining a competence in worldly things? Can it be religion to regard as blessings those things, those very things, which God expressly numbers amongst his curses? Poverty never finds a place amongst

the blessings promised by God. His blessings are of a directly opposite description; flocks, herds, corn, wine and oil; a smiling land; a rejoicing people; abundance for the body and gladness of the heart: these are the blessings which God promises to the industrious, the sober, the careful, and the upright. Let no man, then, believe that, to be poor and wretched is a mark of God's favour; and let no man remain in that state, if he, by any honest means, can rescue himself from it.

18. Poverty leads to all sorts of evil consequences. Want, horrid want, is the great parent of crime. To have a dutiful family, the father's principle of rule must be love not fear. His sway must be gentle, or he will have only an unwilling and short-lived obedience. But it is given to but few men to be gentle and good-humoured amidst the various torments attendant on pinching poverty. A competence is, therefore, the first thing to be thought of; it is the foundation of all good in the labourer's dwelling; without it little but misery can be expected. "Health, peace, and competence," one of the wisest of men regards as the only things needful to man: but the two former are scarcely to be had without the latter. Competence is the foundation of happiness and of exertion. Beset with wants, having a mind continually harassed with fears of starvation, who can act with energy, who can calmly think? To provide a good living, therefore, for himself and family, is the very first duty of every man. "Two things," says AGUR, "have I asked; deny me them not before I die: remove far from me vanity and lies; give me neither poverty nor riches; feed me with food convenient for me: lest I be full and deny thee; or lest I be poor and steal."

19. A good living therefore, a competence, is the first thing to be desired and to be sought after; and, if this little work should have the effect of aiding only a small portion of the Labouring Classes in securing that competence, it will afford great gratification to their friend

WM. COBBETT.

Kensington, 19th July, 1821.

BREWING BEER.

20. Before I proceed to give any directions about brewing, let me mention some of the inducements to do the thing. In former times, to set about to show to Englishmen that it was good for them to brew beer in their houses would have been as impertinent as gravely to insist, that they ought to endeavour not to lose their breath; for, in those times, (only forty years ago,) to have a house and not to brew was a rare thing indeed. Mr. ELLMAN, an old man and a large farmer, in Sussex, has recently given in evidence, before a Committee of the House of Commons, this fact; that, forty years ago, there was not a labourer in his parish that did not brew his own beer; and that now there is not one that does it, except by chance the malt be given him. The causes of this change have been the lowering of the wages of labour, compared with the price of provisions, by the means of the paper-money; the enormous tax upon the barley when made into malt; and the increased tax upon hops. These have quite changed the customs of the English people as to their drink. They still drink beer, but, in general, it is of the brewing of common brewers, and in public-houses, of which the common brewers have become the owners, and have thus, by the aid of paper-money, obtained a monopoly in the supplying of the great body of the people with one of those things which, to the hard-working man, is almost a necessary of life.

21. These things will be altered. They must be altered. The nation must be sunk into nothingness, or a new system must be adopted; and the nation will not sink into nothingness. The malt now pays a tax of 4s. 6d.[1] a bushel, and the barley costs only 3s. This brings the bushel of malt to 8s. including the maltster's charge for malting. If the tax were taken off the malt, malt would be sold, at the present price of barley, for about 3s. 3d. a bushel; because a bushel of barley makes more than a bushel of malt, and the tax, besides its amount, causes great expenses of various sorts to the maltster. The hops pay a tax of 2d.[2] a pound; and a bushel of malt requires, in general, a pound of hops; if these two taxes were taken off, therefore, the consumption of barley and of hops would be exceedingly increased; for double the present quantity would be demanded, and the land is always ready to send it forth.

22. It appears impossible that the landlords should much longer submit to these intolerable burdens on their estates. In short, they must get off the malt tax, or lose those estates. They must do a great deal more, indeed; but that they must do at any rate. The paper-money is fast losing its destructive power; and things are, with regard to the labourers, coming back to what they were forty years ago, and therefore we may prepare for the making of beer in our own houses, and take leave of the poisonous stuff served out to us by common brewers. We may begin immediately; for, even at present prices, home-brewed beer is the cheapest drink that a family can use, except milk, and milk can be applicable only in certain cases.

23. The drink which has come to supply the place of beer has, in general, been tea. It is notorious that tea has no useful strength in it; that it contains nothing nutritious; that it, besides being good for nothing, has badness in it, because it is well known to produce want of sleep in many cases, and in all cases, to shake and weaken the nerves. It is, in fact, a weaker kind of

laudanum, which enlivens for the moment and deadens afterwards. At any rate it communicates no strength to the body; it does not, in any degree, assist in affording what labour demands. It is, then, of no use. And, now, as to its cost, compared with that of beer. I shall make my comparison applicable to a year, or three hundred and sixty-five days. I shall suppose the tea to be only five shillings the pound; the sugar only sevenpence; the milk only twopence a quart. The prices are at the very lowest. I shall suppose a tea-pot to cost a shilling, six cups and saucers two shillings and sixpence, and six pewter spoons eighteen-pence. How to estimate the firing I hardly know; but certainly there must be in the course of the year, two hundred fires made that would not be made, were it not for tea drinking.

Then comes the great article of all, the time employed in this tea-making affair. It is impossible to make a fire, boil water, make the tea, drink it, wash up the things, sweep up the fire-place, and put all to rights again, in a less space of time, upon an average, than two hours. However, let us allow one hour; and here we have a woman occupied no less than three hundred and sixty-five hours in the year, or thirty whole days, at twelve hours in the day; that is to say, one month out of the twelve in the year, besides the waste of the man's time in hanging about waiting for the tea! Needs there any thing more to make us cease to wonder at seeing labourers' children with dirty linen and holes in the heels of their stockings? Observe, too, that the time thus spent is, one half of it, the best time of the day. It is the top of the morning, which, in every calling of life, contains an hour worth two or three hours of the afternoon. By the time that the clattering tea tackle is out of the way, the morning is spoiled; its prime is gone; and any work that is to be done afterwards lags heavily along. If the mother have to go out to work, the tea affair must all first be over. She comes into the field, in summer time, when the sun has gone a third part of his course. She has the heat of the day to encounter, instead of having her work done and being ready to return home at any early hour. Yet early she must go, too: for, there is the fire again to be made, the clattering tea-tackle again to come forward; and even in the longest day she must have candle light, which never ought to be seen in a cottage (except in case of illness) from March to September.

24. Now, then, let us take the bare cost of the use of tea. I suppose a pound of tea to last twenty days; which is not nearly half an ounce every morning and evening. I allow for each mess half a pint of milk. And I allow three pounds of the red dirty sugar to each pound of tea. The account of expenditure would then stand very high; but to these must be added the amount of the tea tackle, one set of which will, upon an average, be demolished every year. To these outgoings must be added the cost of beer at the public-house; for some the man will have, after all, and the woman too, unless they be upon the point of actual starvation. Two pots a week is as little as will serve in this way; and here is a dead loss of ninepence a week, seeing that two pots of beer, full as strong, and a great deal better, can be brewed at home for threepence. The account of the year's tea drinking will then stand thus:

L. s. d.

18lb. of tea 4 10 0 54lb. of sugar 1 11 6 365 pints of milk 1 10 0 Tea tackle 0 5 0 200 fires 0 16 8 30 days' work 0 15 0 Loss by going to public-house 1 19 0 ----------- L. 11 7 2

25. I have here estimated every thing at its very lowest. The entertainment which I have here provided is as poor, as mean, as miserable as any thing short of starvation can set forth; and

yet the wretched thing amounts to a good third part of a good and able labourer's wages! For this money, he and his family may drink good and wholesome beer; in a short time, out of the mere savings from this waste, may drink it out of silver cups and tankards. In a labourer's family, wholesome beer, that has a little life in it, is all that is wanted in general. Little children, that do not work, should not have beer. Broth, porridge, or something in that way, is the thing for them. However, I shall suppose, in order to make my comparison as little complicated as possible, that he brews nothing but beer as strong as the generality of beer to be had at the public-house, and divested of the poisonous drugs which that beer but too often contains; and I shall further suppose that he uses in his family two quarts of this beer every day from the first of October to the last day of March inclusive: three quarts a day during the months of April and May; four quarts a day during the months of June and September; and five quarts a day during the months of July and August; and if this be not enough, it must be a family of drunkards. Here are 1097 quarts, or 274 gallons. Now, a bushel of malt will make eighteen gallons of better beer than that which is sold at the public-houses. And this is precisely a gallon for the price of a quart. People should bear in mind, that the beer bought at the public-house is loaded with a beer tax, with the tax on the public-house keeper, in the shape of license, with all the taxes and expenses of the brewer, with all the taxes, rent, and other expenses of the publican, and with all the profits of both brewer and publican; so that when a man swallows a pot of beer at a public-house, he has all these expenses to help to defray, besides the mere tax on the malt and on the hops.

26. Well, then, to brew this ample supply of good beer for a labourer's family, these 274 gallons, requires fifteen bushels of malt and (for let us do the thing well) fifteen pounds of hops. The malt is now eight shillings a bushel, and very good hops may be bought for less than a shilling a pound. The grains and yeast will amply pay for the labour and fuel employed in the brewing; seeing that there will be pigs to eat the grains, and bread to be baked with the yeast. The account will then stand thus:

L. s. d. 15 bushels of malt 6 0 0 15 pounds of hops 0 15 0 Wear of utensils 0 10 0 ----------- L. 7 5 0

27. Here, then, is the sum of four pounds two shillings and twopence saved every year. The utensils for brewing are, a brass kettle, a mashing tub, coolers, (for which washing tubs may serve,) a half hogshead, with one end taken out, for a tun tub, about four nine-gallon casks, and a couple of eighteen-gallon casks. This is an ample supply of utensils, each of which will last, with proper care, a good long lifetime or two, and the whole of which, even if purchased new from the shop, will only exceed by a few shillings, if they exceed at all, the amount of the saving, arising the very first year, from quitting the troublesome and pernicious practice of drinking tea. The saving of each succeeding year would, if you chose it, purchase a silver mug to hold half a pint at least. However, the saving would naturally be applied to purposes more conducive to the well-being and happiness of a family.

28. It is not, however, the mere saving to which I look. This is, indeed, a matter of great importance, whether we look at the amount itself, or at the ultimate consequences of a judicious application of it; for four pounds make a great hole in a man's wages for the year; and when we consider all the advantages that would arise to a family of children from having these four pounds, now so miserably wasted, laid out upon their backs, in the shape of a decent dress, it is

impossible to look at this waste without feelings of sorrow not wholly unmixed with those of a harsher description.

29. But, I look upon the thing in a still more serious light. I view the tea drinking as a destroyer of health, an enfeebler of the frame, an engenderer of effeminacy and laziness, a debaucher of youth, and a maker of misery for old age. In the fifteen bushels of malt there are 570 pounds weight of sweet; that is to say, of nutricious matter, unmixed with any thing injurious to health. In the 730 tea messes of the year there are 54 pounds of sweet in the sugar, and about 30 pounds of matter equal to sugar in the milk. Here are 84 pounds instead of 570, and even the good effect of these 84 pounds is more than over-balanced by the corrosive, gnawing and poisonous powers of the tea.

30. It is impossible for any one to deny the truth of this statement. Put it to the test with a lean hog: give him the fifteen bushels of malt, and he will repay you in ten score of bacon or thereabouts. But give him the 730 tea messes, or rather begin to give them to him, and give him nothing else, and he is dead with hunger, and bequeaths you his skeleton, at the end of about seven days. It is impossible to doubt in such a case. The tea drinking has done a great deal in bringing this nation into the state of misery in which it now is; and the tea drinking, which is carried on by "dribs" and "drabs;" by pence and farthings going out at a time; this miserable practice has been gradually introduced by the growing weight of the taxes on malt and on hops, and by the everlasting penury amongst the labourers, occasioned by the paper-money.

31. We see better prospects however, and therefore let us now rouse ourselves, and shake from us the degrading curse, the effects of which have been much more extensive and infinitely more mischievous than men in general seem to imagine.

32. It must be evident to every one, that the practice of tea drinking must render the frame feeble and unfit to encounter hard labour or severe weather, while, as I have shown, it deducts from the means of replenishing the belly and covering the back. Hence succeeds a softness, an effeminacy, a seeking for the fire-side, a lurking in the bed, and, in short, all the characteristics of idleness, for which, in this case, real want of strength furnishes an apology. The tea drinking fills the public-house, makes the frequenting of it habitual, corrupts boys as soon as they are able to move from home, and does little less for the girls, to whom the gossip of the tea-table is no bad preparatory school for the brothel. At the very least, it teaches them idleness. The everlasting dawdling about with the slops of the tea tackle, gives them a relish for nothing that requires strength and activity. When they go from home, they know how to do nothing that is useful. To brew, to bake, to make butter, to milk, to rear poultry; to do any earthly thing of use they are wholly unqualified. To shut poor young creatures up in manufactories is bad enough; but there, at any rate, they do something that is useful; whereas, the girl that has been brought up merely to boil the tea-kettle, and to assist in the gossip inseparable from the practice, is a mere consumer of food, a pest to her employer, and a curse to her husband, if any man be so unfortunate as to fix his affections upon her.

33. But is it in the power of any man, any good labourer, who has attained the age of fifty, to look back upon the last thirty years of his life, without cursing the day in which tea was introduced into England? Where is there such a man, who cannot trace to this cause a very considerable part of all the mortifications and sufferings of his life? When was he ever too late

at his labour; when did he ever meet with a frown, with a turning off, and pauperism on that account, without being able to trace it to the tea-kettle? When reproached with lagging in the morning, the poor wretch tells you that he will make up for it by working during his breakfast time! I have heard this a hundred and a hundred times over. He was up time enough; but the tea-kettle kept him lolling and lounging at home; and now, instead of sitting down to a breakfast upon bread, bacon, and beer, which is to carry him on to the hour of dinner, he has to force his limbs along under the sweat of feebleness, and at dinner time to swallow his dry bread, or slake his half-feverish thirst at the pump or the brook. To the wretched tea-kettle he has to return at night, with legs hardly sufficient to maintain him; and thus he makes his miserable progress towards that death, which he finds ten or fifteen years sooner than he would have found it had he made his wife brew beer instead of making tea. If he now and then gladdens his heart with the drugs of the public house, some quarrel, some accident, some illness, is the probable consequence; to the affray abroad succeeds an affray at home; the mischievous example reaches the children, corrupts them or scatters them, and misery for life is the consequence.

34. I should now proceed to the details of brewing; but these, though they will not occupy a large space, must be put off to the second number. The custom of brewing at home has so long ceased amongst labourers, and, in many cases, amongst tradesmen, that it was necessary for me fully to state my reasons for wishing to see the custom revived. I shall, in my next, clearly explain how the operation is performed; and it will be found to be so easy a thing, that I am not without hope, that many tradesmen, who now spend their evenings at the public house, amidst tobacco smoke and empty noise, may be induced, by the finding of better drink at home, at a quarter part of the price, to perceive that home is by far the pleasantest place wherein to pass their hours of relaxation.

35. My work is intended chiefly for the benefit of cottagers, who must, of course, have some land; for, I purpose to show, that a large part of the food of even a large family may be raised, without any diminution of the labourer's earnings abroad, from forty rod, or a quarter of an acre, of ground; but at the same time, what I have to say will be applicable to larger establishments, in all the branches of domestic economy: and especially to that of providing a family with beer.

36. The kind of beer, for a labourer's family, that is to say, the degree of strength, must depend on circumstances; on the numerousness of the family; on the season of the year, and various other things. But, generally speaking, beer half the strength of that mentioned in paragraph 25 will be quite strong enough; for that is, at least, one-third stronger than the farm-house "small beer," which, however, as long experience has proved, is best suited to the purpose. A judicious labourer would probably always have some ale in his house, and have small beer for the general drink. There is no reason why he should not keep Christmas as well as the farmer; and when he is mowing, reaping, or is at any other hard work, a quart, or three pints, of really good fat ale a-day is by no means too much. However, circumstances vary so much with different labourers, that as to the sort of beer, and the number of brewings, and the times of brewing, no general rule can be laid down.

37. Before I proceed to explain the uses of the several brewing utensils, I must speak of the quality of the materials of which beer is made; that is to say, the malt, hops, and water. Malt varies very much in quality, as, indeed, it must, with the quality of the barley. When good, it is

full of flour, and in biting a grain asunder, you find it bite easily, and see the shell thin and filled up well with flour. If it bite hard and steely, the malt is bad. There is pale malt and brown malt; but the difference in the two arises merely from the different degrees of heat employed in the drying. The main thing to attend to is, the quantity of flour. If the barley was bad; thin, or steely, whether from unripeness or blight, or any other cause, it will not malt so well; that is to say, it will not send out its roots in due time; and a part of it will still be barley. Then, the world is wicked enough to think, and even to say, that there are maltsters who, when they send you a bushel of malt, put a little barley amongst it, the malt being taxed and the barley not! Let us hope that this is seldom the case; yet, when we do know that this terrible system of taxation induces the beer-selling gentry to supply their customers with stuff little better than poison, it is not very uncharitable to suppose it possible for some maltsters to yield to the temptations of the devil so far as to play the trick above mentioned. To detect this trick, and to discover what portion of the barley is in an unmalted state, take a handful of the unground malt, and put it into a bowl of cold water. Mix it about with the water a little; that is, let every grain be just wet all over; and whatever part of them sink are not good. If you have your malt ground, there is not, as I know of, any means of detection. Therefore, if your brewing be considerable in amount, grind your own malt, the means of doing which is very easy, and neither expensive nor troublesome, as will appear, when I come to speak of flour. If the barley be well malted, there is still a variety in the quality of the malt; that is to say, a bushel of malt from fine, plump, heavy barley, will be better than the same quantity from thin and light barley. In this case, as in the case of wheat, the weight is the criterion of the quality. Only bear in mind, that as a bushel of wheat, weighing sixty-two pounds, is better worth six shillings, than a bushel weighing fifty-two is worth four shillings, so a bushel of malt weighing forty-five pounds is better worth nine shillings, than a bushel weighing thirty-five is worth six shillings. In malt, therefore, as in every thing else, the word cheap is a deception, unless the quality be taken into view. But, bear in mind, that in the case of unmalted barley, mixed with the malt, the weight can be no rule; for barley is heavier than malt.

NO. II.

BREWING BEER--(CONTINUED.)

38. As to using barley in the making of beer, I have given it a full and fair trial twice over, and I would recommend it to neither rich nor poor. The barley produces strength, though nothing like the malt; but the beer is flat, even though you use half malt and half barley; and flat beer lies heavy on the stomach, and of course, besides the bad taste, is unwholesome. To pay 4s. 6d. tax upon every bushel of our own barley, turned into malt, when the barley itself is not worth 3s. a bushel, is a horrid thing; but, as long as the owners of the land shall be so dastardly as to suffer themselves to be thus deprived of the use of their estates to favour the slave-drivers and plunderers of the East and West Indies, we must submit to the thing, incomprehensible to foreigners, and even to ourselves, as the submission may be.

39. With regard to hops, the quality is very various. At times when some sell for 5s. a pound, others sell for sixpence. Provided the purchaser understand the article, the quality is, of course, in proportion to the price. There are two things to be considered in hops: the power of preserving beer, and that of giving it a pleasant flavour. Hops may be strong; and yet not good. They should be bright, have no leaves or bits of branches amongst them. The hop is the husk, or seed-pod, of the hop-vine, as the cone is that of the fir-tree; and the seeds themselves are deposited, like those of the fir, round a little soft stalk, enveloped by the several folds of this pod, or cone. If, in the gathering, leaves of the vine or bits of the branches are mixed with the hops, these not only help to make up the weight, but they give a bad taste to the beer; and indeed, if they abound much, they spoil the beer. Great attention is therefore necessary in this respect. There are, too, numerous sorts of hops, varying in size, form, and quality, quite as much as apples. However, when they are in a state to be used in brewing, the marks of goodness are an absence of brown colour, (for that indicates perished hops;) a colour between green and yellow; a great quantity of the yellow farina; seeds not too large nor too hard; a clammy feel when rubbed between the fingers; and a lively, pleasant smell. As to the age of hops, they retain for twenty years, probably, their power of preserving beer; but not of giving it a pleasant flavour. I have used them at ten years old, and should have no fear of using them at twenty. They lose none of their bitterness; none of their power of preserving beer; but they lose the other quality; and therefore, in the making of fine ale, or beer, new hops are to be preferred. As to the quantity of hops, it is clear, from what has been said, that that must, in some degree depend upon their quality; but, supposing them to be good in quality, a pound of hops to a bushel of malt is about the quantity. A good deal, however, depends upon the length of time that the beer is intended to be kept, and upon the season of the year in which it is brewed. Beer intended to be kept a long while should have the full pound, also beer brewed in warmer weather, though for present use: half the quantity may do under an opposite state of circumstances.

40. The water should be soft by all means. That of brooks, or rivers, is best. That of a pond, fed by a rivulet, or spring, will do very well. Rain-water, if just fallen, may do; but stale rain-water, or stagnant pond-water, makes the beer flat and difficult to keep; and hard water, from wells, is very bad; it does not get the sweetness out of the malt, nor the bitterness out of the hops, like soft water; and the wort of it does not ferment well, which is a certain proof of its unfitness for the purpose.

41. There are two descriptions of persons whom I am desirous to see brewing their own beer; namely, tradesmen, and labourers and journeymen. There must, therefore, be two distinct scales treated of. In the former editions of this work, I spoke of a machine for brewing, and stated the advantages of using it in a family of any considerable consumption of beer; but, while, from my desire to promote private brewing, I strongly recommended the machine, I stated that, "if any of my readers could point out any method by which we should be more likely to restore the practice of private brewing, and especially to the cottage, I should be greatly obliged to them to communicate it to me." Such communications have been made, and I am very happy to be able, in this new edition of my little work, to avail myself of them. There was, in the Patent Machine, always, an objection on account of the expense; for, even the machine for one bushel of malt cost, at the reduced price, eight pounds; a sum far above the reach of a cottager, and even above that of a small tradesman. Its convenience, especially in towns, where room is so valuable, was an object of great importance; but there were disadvantages attending it which, until after some experience, I did not ascertain. It will be remembered that the method by the brewing machine requires the malt to be put into the cold water, and for the water to make the malt swim, or, at least, to be in such proportion as to prevent the fire beneath from burning the malt. We found that our beer was flat, and that it did not keep. And this arose, I have every reason to believe, from this process. The malt should be put into hot water, and the water, at first, should be but just sufficient in quantity to stir the malt in, and separate it well. Nevertheless, when it is merely to make small beer; beer not wanted to keep; in such cases the brewing machine may be of use; and, as will be seen by-and-by, a moveable boiler (which has nothing to do with the patent) may, in many cases, be of great convenience and utility.

42. The two scales of which I have spoken above, are now to be spoken of; and, that I may explain my meaning the more clearly, I shall suppose, that, for the tradesman's family, it will be requisite to brew eighteen gallons of ale and thirty-six of small beer, to fill three casks of eighteen gallons each. It will be observed, of course, that, for larger quantities, larger utensils of all sorts will be wanted. I take this quantity as the one to give directions on. The utensils wanted here will be, FIRST, a copper that will contain forty gallons, at least; for, though there be to be but thirty-six gallons of small beer, there must be space for the hops, and for the liquor that goes off in steam. SECOND, a mashing-tub to contain sixty gallons; for the malt is to be in this along with the water. THIRD, an underbuck, or shallow tub to go under the mash-tub, for the wort to run into when drawn from the grains.

FOURTH, a tun-tub, that will contain thirty gallons, to put the ale into to work, the mash-tub, as we shall see, serving as a tun-tub for the small beer. Besides these, a couple of coolers, shallow tubs, which may be the heads of wine buts, or some such things, about a foot deep; or if you have four it may be as well, in order to effect the cooling more quickly.

43. You begin by filling the copper with water, and next by making the water boil. You then put into the mashing-tub water sufficient to stir and separate the malt in. But now let me say more particularly what this mashing-tub is. It is, you know, to contain sixty gallons. It is to be a little broader at top than at bottom, and not quite so deep as it is wide across the bottom. Into the middle of the bottom there is a hole about two inches over, to draw the wort off through. In this hole goes a stick, a foot or two longer than the tub is high. This stick is to be about two inches through, and tapered for about eight inches upwards at the end that goes into the hole, which at last it fills up closely as a cork. Upon the hole, before any thing else be put into the tub, you lay a little bundle of fine birch, (heath or straw may do,) about half the bulk of a birch broom, and well tied at both ends. This being laid over the hole (to keep back the grains as the wort goes out,) you put the tapered end of the stick down through into the hole, and thus cork the hole up. You must then have something of weight sufficient to keep the birch steady at the bottom of the tub, with a hole through it to slip down the stick; otherwise when the stick is raised it will be apt to raise the birch with it, and when you are stirring the mash you would move it from its place. The best thing for this purpose will be a leaden collar for the stick, with the hole in the collar plenty large enough, and it should weigh three or four pounds. The thing they use in some farm-houses is the iron box of a wheel. Any thing will do that will slide down the stick, and lie with weight enough on the birch to keep it from moving. Now, then, you are ready to begin brewing. I allow two bushels of malt for the brewing I have supposed. You must now put into the mashing-tub as much boiling water as will be sufficient to stir the malt in and separate it well. But here occur some of the nicest points of all; namely, the degree of heat that the water is to be at, before you put in the malt. This heat is one hundred and seventy degrees by the thermometer. If you have a thermometer, this is ascertained easily; but, without one, take this rule, by which so much good beer has been made in England for hundreds of years: when you can, by looking down into the tub, see your face clearly in the water, the water is become cool enough; and you must not put the malt in before. Now put in the malt and stir it well in the water. To perform this stirring, which is very necessary, you have a stick, somewhat bigger than a broom-stick, with two or three smaller sticks, eight or ten inches long, put through the lower end of it at about three or four inches asunder, and sticking out on each side of the long stick. These small cross sticks serve to search the malt and separate it well in the stirring or mashing. Thus, then, the malt is in; and in this state it should continue for about a quarter of an hour. In the mean while you will have filled up your copper, and made it boil; and now (at the end of the quarter of an hour) you put in boiling water sufficient to give you your eighteen gallons of ale. But, perhaps, you must have thirty gallons of water in the whole; for the grains will retain at least ten gallons of water; and it is better to have rather too much wort than too little. When your proper quantity of water is in, stir the malt again well. Cover the mashing-tub over with sacks, or something that will answer the same purpose; and there let the mash stand for two hours. When it has stood the two hours, you draw off the wort. And now, mind, the mashing-tub is placed on a couple of stools, or on something, that will enable you to put the underbuck under it, so as to receive the wort as it comes out of the hole before-mentioned. When you have put the underbuck in its place, you let out the wort by pulling up the stick that corks the whole. But, observe, this stick (which goes six or eight inches through the hole) must be raised by degrees, and the wort must be let out slowly, in order to keep back the sediment. So that it is necessary to have something to keep the stick up at the point where you are to raise it, and wish to fix it at for the time. To do this, the simplest, cheapest and best thing in the world is a cleft stick. Take a rod of ash, hazel, birch, or almost any wood; let it be a foot or two longer than your mashing-tub is wide over the

top; split it, as if for making hoops; tie it round with a string at each end; lay it across your mashing-tub; pull it open in the middle, and let the upper part of the wort-stick through it, and when you raise that stick, by degrees as before directed, the cleft stick will hold it up at whatever height you please.

44. When you have drawn off the ale-wort, you proceed to put into the mashing tub water for the small beer. But, I shall go on with my directions about the ale till I have got it into the cask and cellar; and shall then return to the small-beer.

45. As you draw off the ale-wort into the underbuck, you must lade it out of that into the tun-tub, for which work, as well as for various other purposes in the brewing, you must have a bowl-dish with a handle to it. The underbuck will not hold the whole of the wort. It is, as before described, a shallow tub, to go under the mashing-tub to draw off the wort into. Out of this underbuck you must lade the ale-wort into the tun-tub; and there it must remain till your copper be emptied and ready to receive it.

46. The copper being empty, you put the wort into it, and put in after the wort, or before it, a pound and a half of good hops, well rubbed and separated as you put them in. You now make the copper boil, and keep it, with the lid off, at a good brisk boil, for a full hour, and if it be an hour and a half it is none the worse.

47. When the boiling is done, put out your fire, and put the liquor into the coolers. But it must be put into the coolers without the hops. Therefore, in order to get the hops out of the liquor, you must have a strainer. The best for your purpose is a small clothes-basket, or any other wicker-basket. You set your coolers in the most convenient place. It may be in-doors or out of doors, as most convenient. You lay a couple of sticks across one of the coolers, and put the basket upon them. Put your liquor, hops and all, into the basket, which will keep back the hops. When you have got liquor enough in one cooler, you go to another with your sticks and basket, till you have got all your liquor out. If you find your liquor deeper in one cooler than the other, you can make an alteration in that respect, till you have the liquor so distributed as to cool equally fast in both, or all, the coolers.

48. The next stage of the liquor is in the tun-tub, where it is set to work. Now, a very great point is, the degree of heat that the liquor is to be at when it is set to working. The proper heat is seventy degrees; so that a thermometer makes this matter sure. In the country they determine the degree of heat by merely putting a finger into the liquor. Seventy degrees is but just warm, a gentle luke-warmth. Nothing like heat. A little experience makes perfectness in such a matter. When at the proper heat, or nearly, (for the liquor will cool a little in being removed,) put it into the tun-tub. And now, before I speak of the act of setting the beer to work, I must describe this tun-tub, which I first mentioned in Paragraph 42. It is to hold thirty gallons, as you have seen; and nothing is better than an old cask of that size, or somewhat larger, with the head taken out, or cut off. But, indeed, any tub of sufficient dimensions, and of about the same depth proportioned to the width as a cask or barrel has, will do for the purpose. Having put the liquor into the tun-tub, you put in the yeast. About half a pint of good yeast is sufficient. This should first be put into a thing of some sort that will hold about a gallon of your liquor; the thing should then be nearly filled with liquor, and with a stick or spoon you should mix the yeast well with the liquor in this bowl, or other thing, and stir in along with the yeast a handful of wheat or rye flour.

This mixture is then to be poured out clean into the tun-tub, and the whole mass of the liquor is then to be agitated well by lading up and pouring down again with your bowl-dish, till the yeast be well mixed with the liquor. Some people do the thing in another manner.

They mix up the yeast and flour with some liquor (as just mentioned) taken out of the coolers; and then they set the little vessel that contains this mixture down on the bottom of the tun-tub; and, leaving it there, put the liquor out of the coolers into the tun-tub. Being placed at the bottom, and having the liquor poured on it, the mixture is, perhaps, more perfectly effected in this way than in any way. The flour may not be necessary; but, as the country people use it, it is, doubtless, of some use; for their hereditary experience has not been for nothing. When your liquor is thus properly put into the tun-tub and set a working, cover over the top of the tub by laying across it a sack or two, or something that will answer the purpose.

49. We now come to the last stage; the cask or barrel. But I must first speak of the place for the tun-tub to stand in. The place should be such as to avoid too much warmth or cold. The air should, if possible, be at about 55 degrees. Any cool place in summer and any warmish place in winter. If the weather be very cold, some cloths or sacks should be put round the tun-tub while the beer is working. In about six or eight hours, a frothy head will rise upon the liquor; and it will keep rising, more or less slowly, for about forty-eight hours. But, the length of time required for the working depends on various circumstances; so that no precise time can be fixed. The best way is, to take off the froth (which is indeed yeast) at the end of about twenty-four hours, with a common skimmer, and put it into a pan or vessel of some sort; then, in twelve hours' time, take it off again in the same way; and so on till the liquor has done working, and sends up no more yeast.

Then it is beer; and when it is quite cold (for ale or strong beer) put it into the cask by means of a funnel. It must be cold before you do this, or it will be what the country-people call foxed; that is to say, have a rank and disagreeable taste. Now, as to the cask, it must be sound and sweet. I thought, when writing the former edition of this work, that the bell-shaped were the best casks. I am now convinced that that was an error. The bell-shaped, by contracting the width of the top of the beer, as that top descends, in consequence of the draft for use, certainly prevents the head (which always gathers on beer as soon as you begin to draw it off) from breaking and mixing in amongst the beer. This is an advantage in the bell-shape; but then the bell-shape, which places the widest end of the cask uppermost, exposes the cask to the admission of external air much more than the other shape. This danger approaches from the ends of the cask; and, in the bell-shape, you have the broadest end wholly exposed the moment you have drawn out the first gallon of beer, which is not the case with the casks of the common shape. Directions are given, in the case of the bell-casks, to put damp sand on the top to keep out the air. But, it is very difficult to make this effectual; and yet, if you do not keep out the air, your beer will be flat; and when flat, it really is good for nothing but the pigs. It is very difficult to fill the bell-cask, which you will easily see if you consider its shape. It must be placed on the level with the greatest possible truth, or there will be a space left; and to place it with such truth is, perhaps, as difficult a thing as a mason or bricklayer ever had to perform.

And yet, if this be not done, there will be an empty space in the cask, though it may, at the same time, run over. With the common casks there are none of these difficulties. A common eye will see when it is well placed; and, at any rate, any little vacant space that may be left is not at an end of the cask, and will, without great carelessness, be so small as to be of no consequence. We now come to the act of putting in the beer. The cask should be placed on a stand with legs about a foot long. The cask, being round, must have a little wedge, or block, on each side to keep it steady. Bricks do very well. Bring your beer down into the cellar in buckets, and pour it in through the funnel, until the cask be full. The cask should lean a little on one side, when you fill it; because the beer will work again here, and send more yeast out of the bung-hole; and, if the cask were not a little on one side, the yeast would flow over both sides of the cask, and would not descend in one stream into a pan, put underneath to receive it. Here the bell-cask is extremely inconvenient; for the yeast works up all over the head, and cannot run off, and makes a very nasty affair. This alone, to say nothing of the other disadvantages, would decide the question against the bell-casks. Something will go off in this working, which may continue for two or three days. When you put the beer in the cask, you should have a gallon or two left, to keep filling up with as the working produces emptiness. At last, when the working is completely over, right the cask.

That is to say, block it up to its level. Put in a handful of fresh hops. Fill the cask quite full. Put in the bung, with a bit of coarse linen stuff round it; hammer it down tight; and, if you like, fill a coarse bag with sand, and lay it, well pressed down, over the bung.

50. As to the length of time that you are to keep the beer before you begin to use it, that must, in some measure, depend on taste. Such beer as this ale will keep almost any length of time. As to the mode of tapping, that is as easy almost as drinking. When the cask is empty, great care must be taken to cork it tightly up, so that no air get in; for, if it do, the cask is moulded, and when once moulded, it is spoiled for ever. It is never again fit to be used about beer. Before the cask be used again, the grounds must be poured out, and the cask cleaned by several times scalding; by putting in stones (or a chain,) and rolling and shaking about till it be quite clean. Here again the round casks have the decided advantage; it being almost impossible to make the bell-casks thoroughly clean, without taking the head out, which is both troublesome and expensive; as it cannot be well done by any one but a cooper, who is not always at hand, and who, when he is, must be paid.

51. I have now done with the ale, and it remains for me to speak of the small beer. In Paragraph 47 (which now see) I left you drawing off the ale-wort, and with your copper full of boiling water. Thirty-six gallons of that boiling water are, as soon as you have got your ale-wort out, and have put down your mash-tub stick to close up the hole at the bottom; as soon as you have done this, thirty-six gallons of the boiling water are to go into the mashing-tub; the grains are to be well stirred up, as before; the mashing-tub is to be covered over again, as mentioned in Paragraph 43; and the mash is to stand in that state for an hour, and not two hours, as for the ale-wort.

52. When the small beer mash has stood its hour, draw it off as in Paragraph 47, and put it into the tun-tub as you did the ale-wort.

53. By this time your copper will be empty again, by putting your ale-liquor to cool, as mentioned in Paragraph 47. And you now put the small beer wort into the copper, with the hops that you used before, and with half a pound of fresh hops added to them; and this liquor you boil briskly for an hour.

54. By this time you will have taken the grains and the sediment clean out of the mashing-tub, and taken out the bunch of birch twigs, and made all clean. Now put in the birch twigs again, and put down your stick as before. Lay your two or three sticks across the mashing-tub, put your basket on them, and take your liquor from the copper (putting the fire out first) and pour it into the mashing-tub through the basket. Take the basket away, throw the hops to the dunghill, and leave the small beer liquid to cool in the mashing-tub.

55. Here it is to remain to be set to working as mentioned for the ale, in Paragraph 48; only, in this case, you will want more yeast in proportion; and should have for your thirty-six gallons of small beer, three half pints of good yeast.

56. Proceed, as to all the rest of the business, as with the ale, only, in the case of the small beer, it should be put into the cask, not quite cold, but a little warm; or else it will not work at all in the barrel, which it ought to do.

It will not work so strongly or so long as the ale; and may be put in the barrel much sooner; in general the next day after it is brewed.

57. All the utensils should be well cleaned and put away as soon as they are done with; the little things as well as the great things; for it is loss of time to make new ones. And, now, let us see the expense of these utensils. The copper, new, 5l.; the mashing-tub, new, 30s.; the tun-tub, not new, 5s.; the underbuck and three coolers, not new, 20s. The whole cost is 7l. 10s. which is ten shillings less than the one bushel machine. I am now in a farm-house, where the same set of utensils has been used for forty years; and the owner tells me, that, with the same use, they may last for forty years longer. The machine will not, I think, last four years, if in any thing like regular use. It is of sheet-iron, tinned on the inside, and this tin rusts exceedingly, and is not to be kept clean without such rubbing as must soon take off the tin. The great advantage of the machine is, that it can be removed. You can brew without a brew-house.--You can set the boiler up against any fire-place, or any window. You can brew under a cart-shed, and even out of doors. But all this may be done with these utensils, if your copper be moveable. Make the boiler of copper, and not of sheet-iron, and fix it on a stand with a fire-place and stove-pipe; and then you have the whole to brew out of doors with as well as in-doors, which is a very great convenience.

58. Now with regard to the other scale of brewing, little need be said; because, all the principles being the same, the utensils only are to be proportioned to the quantity. If only one sort of beer be to be brewed at a time, all the difference is, that, in order to extract the whole of the goodness of the malt, the mashing ought to be at twice. The two worts are then put together, and then you boil them together with the hops.

59. A Correspondent at Morpeth says, the whole of the utensils used by him are a twenty-gallon pot, a mashing-tub, that also answers for a tun-tub, and a shallow tub for a cooler; and that these are plenty for a person who is any thing of a contriver. This is very true; and these things will cost no more, perhaps, than forty shillings. A nine gallon cask of beer can be brewed very well with such utensils. Indeed, it is what used to be done by almost every labouring man in the kingdom, until the high price of malt and comparatively low price of wages rendered the people too poor and miserable to be able to brew at all. A Correspondent at Bristol has obligingly

sent me the model of utensils for brewing on a small scale; but as they consist chiefly of brittle ware, I am of opinion that they would not so well answer the purpose.

60. Indeed, as to the country labourers, all they want is the ability to get the malt. Mr. ELLMAN, in his evidence before the Agricultural Committee, said, that, when he began farming, forty-five years ago, there was not a labourer's family in the parish that did not brew their own beer and enjoy it by their own fire-sides; and that, now, not one single family did it, from want of ability to get the malt. It is the tax that prevents their getting the malt; for, the barley is cheap enough. The tax causes a monopoly in the hands of the maltsters, who, when the tax is two and sixpence, make the malt, cost 7s. 6d., though the barley cost but 2s. 6d.; and though the malt, tax and all, ought to cost him about 5s. 6d. If the tax were taken off, this pernicious monopoly would be destroyed.

61. The reader will easily see, that, in proportion to the quantity wanted to be brewed must be the size of the utensils; but, I may observe here, that the above utensils are sufficient for three, or even four, bushels of malt, if stronger beer be wanted.

62. When it is necessary, in case of falling short in the quantity wanted to fill up the ale cask, some may be taken from the small beer. But, upon the whole brewing, there ought to be no falling short; because, if the casks be not filled up, the beer will not be good, and certainly will not keep. Great care should be taken as to the cleansing of the casks. They should be made perfectly sweet; or it is impossible to have good beer.

63. The cellar, for beer to keep any length of time, should be cool. Under a hill is the best place for a cellar; but, at any rate, a cellar of good depth, and dry. At certain times of the year, beer that is kept long will ferment. The vent-pegs must, in such cases, be loosened a little, and afterwards fastened.

64. Small beer may be tapped almost directly. It is a sort of joke that it should see a Sunday; but, that it may do before it be two days old. In short, any beer is better than water; but it should have some strength and some weeks of age at any rate.

65. I cannot conclude this Essay without expressing my pleasure, that a law has been recently passed to authorize the general retail of beer. This really seems necessary to prevent the King's subjects from being poisoned. The brewers and porter quacks have carried their tricks to such an extent, that there is no safety for those who drink brewer's beer.

66. The best and most effectual thing is, however, for people to brew their own beer, to enable them and induce them to do which, I have done all that lies in my power. A longer treatise on the subject would have been of no use. These few plain directions will suffice for those who have a disposition to do the thing, and those who have not would remain unmoved by any thing that I could say.

67. There seems to be a great number of things to do in brewing, but the greater part of them require only about a minute each. A brewing, such as I have given the detail of above, may be completed in a day; but, by the word day, I mean to include the morning, beginning at four o'clock.

68. The putting of the beer into barrel is not more than an hour's work for a servant woman, or a tradesman's or a farmer's wife. There is no heavy work, no work too heavy for a woman in any part of the business, otherwise I would not recommend it to be performed by the women, who, though so amiable in themselves, are never quite so amiable as when they are useful; and as to beauty, though men may fall in love with girls at play, there is nothing to make them stand to their love like seeing them at work. In conclusion of these remarks on beer brewing, I once more express my most anxious desire to see abolished for ever the accursed tax on malt, which, I verily believe, has done more harm to the people of England than was ever done to any people by plague, pestilence, famine, and civil war.

69. In Paragraph 76, in Paragraph 108, and perhaps in another place or two (of the last edition,) I spoke of the machine for brewing. The work being stereotyped, it would have been troublesome to alter those paragraphs; but, of course, the public, in reading them, will bear in mind what has been now said relative to the machine. The inventor of that machine deserves great praise for his efforts to promote private brewing; and, as I said before, in certain confined situations, and where the beer is to be merely small beer, and for immediate use, and where time and room are of such importance as to make the cost of the machine comparatively of trifling consideration, the machine may possibly be found to be an useful utensil.

70. Having stated the inducements to the brewing of beer, and given the plainest directions that I was able to give for the doing of the thing, I shall, next, proceed to the subject of bread. But this subject is too large and of too much moment to be treated with brevity, and must, therefore, be put off till my next Number. I cannot, in the mean while, dismiss the subject of brewing beer without once more adverting to its many advantages, as set forth in the foregoing Number of this work.

71. The following instructions for the making of porter, will clearly show what sort of stuff is sold at public-houses in London; and we may pretty fairly suppose that the public-house beer in the country is not superior to it in quality, "A quarter of malt, with these ingredients, will make five barrels of good porter. Take one quarter of high-coloured malt, eight pounds of hops, nine pounds of treacle, eight pounds of colour, eight pounds of sliced liquorice-root, two drams of salt of tartar, two ounces of Spanish-liquorice, and half an ounce of capsicum." The author says, that he merely gives the ingredients, as used by many persons.

72. This extract is taken from a book on brewing, recently published in London. What a curious composition! What a mess of drugs! But, if the brewers openly avow this, what have we to expect from the secret practices of them, and the retailers of the article! When we know, that beer-doctor and brewers'-druggist are professions, practised as openly as those of bug-man and rat-killer, are we simple enough to suppose that the above-named are the only drugs that people swallow in those potions, which they call pots of beer? Indeed, we know the contrary; for scarcely a week passes without witnessing the detection of some greedy wretch, who has used, in making or in doctoring his beer, drugs, forbidden by the law. And, it is not many weeks since one of these was convicted, in the Court of Excise, for using potent and dangerous drugs, by the means of which, and a suitable quantity of water, he made two buts of beer into three. Upon this occasion, it appeared that no less than ninety of these worthies were in the habit of pursuing the same practices. The drugs are not unpleasant to the taste; they sting the palate: they give a present relish: they communicate a momentary exhilaration: but, they give no force to the body, which,

on the contrary, they enfeeble, and, in many instances, with time, destroy; producing diseases from which the drinker would otherwise have been free to the end of his days.

73. But, look again at the receipt for making porter. Here are eight bushels of malt to 180 gallons of beer; that is to say, twenty-five gallons from the bushel. Now the malt is eight shillings a bushel, and eight pounds of the very best hops will cost but a shilling a pound. The malt and hops, then, for the 180 gallons, cost but seventy-two shillings; that is to say, only a little more than fourpence three farthings a gallon, for stuff which is now retailed for sixteen pence a gallon! If this be not an abomination, I should be glad to know what is. Even if the treacle, colour, and the drugs, be included, the cost is not fivepence a gallon; and yet, not content with this enormous extortion, there are wretches who resort to the use of other and pernicious drugs, in order to increase their gains!

74. To provide against this dreadful evil there is, and there can be, no law; for, it is created by the law. The law it is that imposes the enormous tax on the malt and hops; the law it is that imposes the license tax, and places the power of granting the license at the discretion of persons appointed by the government; the law it is that checks, in this way, the private brewing, and that prevents free and fair competition in the selling of beer, and as long as the law does these, it will in vain endeavour to prevent the people from being destroyed by slow poison.

75. Innumerable are the benefits that would arise from a repeal of the taxes on malt and on hops. Tippling-houses might then be shut up with justice and propriety. The labourer, the artisan, the tradesman, the landlord, all would instantly feel the benefit. But the landlord more, perhaps, in this case, than any other member of the community. The four or five pounds a year which the day-labourer now drizzles away in tea-messes, he would divide with the farmer, if he had untaxed beer. His wages would fall, and fall to his advantage too. The fall of wages would be not less than 40l. upon a hundred acres. Thus 40l. would go, in the end, a fourth, perhaps to the farmer, and three-fourths to the landlord. This is the kind of work to reduce poor-rates, and to restore husbandry to prosperity. Undertaken this work must be, and performed too; but whether we shall see this until the estates have passed away from the present race of landlords, is a question which must be referred to time.

76. Surely we may hope, that, when the American farmers shall see this little Essay, they will begin seriously to think of leaving off the use of the liver-burning and palsy-producing spirits. Their climate, indeed, is something: extremely hot in one part of the year, and extremely cold in the other part of it. Nevertheless, they may have, and do have, very good beer if they will. Negligence is the greatest impediment in their way. I like the Americans very much; and that, if there were no other, would be a reason for my not hiding their faults.

NO. III.

MAKING BREAD.

77. Little time need be spent in dwelling on the necessity of this article to all families; though, on account of the modern custom of using potatoes to supply the place of bread, it seems necessary to say a few words here on the subject, which, in another work I have so amply, and, I think, so triumphantly discussed. I am the more disposed to revive the subject for a moment, in this place, from having read, in the evidence recently given before the Agricultural Committee, that many labourers, especially in the West of England, use potatoes instead of bread to a very great extent. And I find, from the same evidence, that it is the custom to allot to labourers "a potatoe ground" in part payment of their wages! This has a tendency to bring English labourers down to the state of the Irish, whose mode of living, as to food, is but one remove from that of the pig, and of the ill-fed pig too.

78. I was, in reading the above-mentioned Evidence, glad to find, that Mr. EDWARD WAKEFIELD, the best informed and most candid of all the witnesses, gave it as his opinion, that the increase which had taken place in the cultivation of potatoes was "injurious to the country;" an opinion which must, I think, be adopted by every one who takes the trouble to reflect a little upon the subject. For leaving out of the question the slovenly and beastly habits engendered amongst the labouring classes by constantly lifting their principal food at once out of the earth to their mouths, by eating without the necessity of any implements other than the hands and the teeth, and by dispensing with everything requiring skill in the preparation of the food, and requiring cleanliness in its consumption or preservation; leaving these out of the question, though they are all matters of great moment, when we consider their effects in the rearing of a family, we shall find, that, in mere quantity of food, that is to say of nourishment, bread is the preferable diet.

79. An acre of land that will produce 300 bushels of potatoes, will produce 32 bushels of wheat. I state this as an average fact, and am not at all afraid of being contradicted by any one well acquainted with husbandry. The potatoes are supposed to be of a good sort, as it is called, and the wheat may be supposed to weigh 60 pounds a bushel. It is a fact clearly established, that, after the water, the stringy substance, and the earth, are taken from the potatoe, there remains only one tenth of the rough raw weight of nutritious matter, or matter which is deemed equally nutritious with bread, and, as the raw potatoes weigh 56lb. a bushel, the acre will yield 1,830lb. of nutritious matter. Now mind, a bushel of wheat, weighing 60lb. will make of household bread (that is to say, taking out only the bran) 65lb. Thus, the acre yields 2,080lb. of bread. As to the expenses, the seed and act of planting are about equal in the two cases. But, while the potatoes must have cultivation during their growth, the wheat needs none; and while the wheat straw is worth from three to five pounds an acre, the haulm of the potatoes is not worth one single truss

of that straw. Then, as to the expense of gathering, housing, and keeping the potatoe crop, it is enormous, besides the risk of loss by frost, which may be safely taken, on an average, at a tenth of the crop. Then comes the expense of cooking. The thirty-two bushels of wheat, supposing a bushel to be baked at a time, (which would be the case in a large family,) would demand thirty-two heatings of the oven. Suppose a bushel of potatoes to be cooked every day in order to supply the place of this bread, then we have nine hundred boilings of the pot, unless cold potatoes be eaten at some of the meals; and, in that case, the diet must be cheering indeed! Think of the labour; think of the time; think of all the peelings and scrapings and washings and messings attending these nine hundred boilings of the pot! For it must be a considerable time before English people can be brought to eat potatoes in the Irish style; that is to say, scratch them out of the earth with their paws, toss them into a pot without washing, and when boiled, turn them out upon a dirty board, and then sit round that board, peel the skin and dirt from one at a time and eat the inside. Mr. Curwen was delighted with "Irish hospitality," because the people there receive no parish relief; upon which I can only say, that I wish him the exclusive benefit of such hospitality.

80. I have here spoken of a large quantity of each of the sorts of food. I will now come to a comparative view, more immediately applicable to a labourer's family. When wheat is ten shillings the bushel, potatoes, bought at best hand, (I am speaking of the country generally,) are about two shillings (English) a bushel. Last spring the average price of wheat might be six and sixpence, (English;) and the average price of potatoes (in small quantities) was about eighteen-pence; though, by the wagon-load, I saw potatoes bought at a shilling (English) a bushel, to give to sheep; then, observe, these were of the coarsest kind, and the farmer had to fetch them at a considerable expense. I think, therefore, that I give the advantage to the potatoes when I say that they sell, upon an average, for full a fifth part as much as the wheat sells for, per bushel, while they contain four pounds less weight than the bushel of wheat; while they yield only five pounds and a half of nutritious matter equal to bread; and while the bushel of wheat will yield sixty-five pounds of bread, besides the ten pounds of bran. Hence it is clear, that, instead of that saving, which is everlastingly dinned in our ears, from the use of potatoes, there is a waste of more than one half; seeing that, when wheat is ten shillings (English) the bushel, you can have sixty-five pounds of bread for the ten shillings; and can have out of potatoes only five pounds and a half of nutritious matter equal to bread for two shillings! (English.) This being the case, I trust that we shall soon hear no more of those savings which the labourer makes by the use of potatoes; I hope we shall, in the words of Dr. DRENNAN, "leave Ireland to her lazy root," if she choose still to adhere to it. It is the root, also, of slovenliness, filth, misery, and slavery; its cultivation has increased in England with the increase of the paupers: both, I thank God, are upon the decline.

Englishmen seem to be upon the return to beer and bread, from water and potatoes: and, therefore, I shall now proceed to offer some observations to the cottager, calculated to induce him to bake his own bread.

81. As I have before stated, sixty pounds of wheat, that is to say, where the Winchester bushel weighs sixty pounds, will make sixty-five pounds of bread, besides the leaving of about ten pounds of bran. This is household bread, made of flour from which the bran only is taken. If you make fine flour, you take out pollard, as they call it, as well as bran, and then you have a

smaller quantity of bread and a greater quantity of offal; but, even of this finer bread, bread equal in fineness to the baker's bread, you get from fifty-eight to fifty-nine pounds out of the bushel of wheat. Now, then, let us see how many quartern loaves you get out of the bushel of wheat, supposing it to be fine flour, in the first place. You get thirteen quartern loaves and a half; these cost you, at the present average price of wheat (seven and sixpence a bushel,) in the first place 7s. 6d.;[5] then 3d. for yeast; then not more than 3d. for grinding; because you have about thirteen pounds of offal, which is worth more than a 1/2d. a pound, while the grinding is 9d. a bushel. Thus, then, the bushel of bread of fifty-nine pounds costs you eight shillings; and it yields you the weight of thirteen and a half quartern loaves: these quartern loaves now (Dec. 1821) sell at Kensington, at the baker's shop, at 1s. 1/2d.; that is to say, the thirteen quartern loaves and a half cost 14s. 7-1/2d. I omitted to mention the salt, which would cost you 4d. more. So that, here is 6s. 3-1/2d. saved upon the baking of a bushel of bread. The baker's quartern loaf is indeed cheaper in the country than at Kensington, by, probably, a penny in the loaf; which would still, however, leave a saving of 5s. upon the bushel of bread. But, besides this, pray think a little of the materials of which the baker's loaf is composed. The alum, the ground potatoes, and other materials; it being a notorious fact, that the bakers, in London at least, have mills wherein to grind their potatoes; so large is the scale upon which they use that material. It is probable, that, out of a bushel of wheat, they make between sixty and seventy pounds of bread, though they have no more flour, and, of course, no more nutritious matter, than you have in your fifty-nine pounds of bread.

But, at the least, supposing their bread to be as good as yours in quality, you have, allowing a shilling for the heating of the oven, a clear 4s. saved upon every bushel of bread. If you consume half a bushel a week, that is to say about a quartern loaf a day, this is a saving of 5l. 4s. a year, or full a sixth part, if not a fifth part, of the earnings of a labourer in husbandry.

82. How wasteful, then, and, indeed, how shameful, for a labourer's wife to go to the baker's shop; and how negligent, how criminally careless of the welfare of his family, must the labourer be, who permits so scandalous a use of the proceeds of his labour! But I have hitherto taken a view of the matter the least possibly advantageous to the home-baked bread. For, ninety-nine times out of a hundred, the fuel for heating the oven costs very little. The hedgers, the copsers, the woodmen of all descriptions, have fuel for little or nothing. At any rate, to heat the oven cannot, upon an average, take the country through, cost the labourer more than 6d. a bushel. Then, again, fine flour need not ever be used, and ought not to be used. This adds six pounds of bread to the bushel, or nearly another quartern loaf and a half, making nearly fifteen quartern loaves out of the bushel of wheat. The finest flour is by no means the most wholesome; and, at any rate, there is more nutritious matter in a pound of household bread than in a pound of baker's bread. Besides this, rye, and even barley, especially when mixed with wheat, make very good bread. Few people upon the face of the earth live better than the Long Islanders. Yet nine families out of ten seldom eat wheaten-bread. Rye is the flour that they principally make use of. Now, rye is seldom more than two-thirds the price of wheat, and barley is seldom more than half the price of wheat. Half rye and half wheat, taking out a little more of the offal, make very good bread. Half wheat, a quarter rye and a quarter barley, nay, one-third of each, make bread that I could be very well content to live upon all my lifetime; and, even barley alone, if the barley be good, and none but the finest flour taken out of it, has in it, measure for measure, ten times the nutrition of potatoes. Indeed the fact is well known, that our forefathers used barley bread to a

very great extent. Its only fault, with those who dislike it, is its sweetness, a fault which we certainly have not to find with the baker's loaf, which has in it little more of the sweetness of grain than is to be found in the offal which comes from the sawings of deal boards. The nutritious nature of barley is amply proved by the effect, and very rapid effect, of its meal, in the fatting of hogs and of poultry of all descriptions. They will fatten quicker upon meal of barley than upon any other thing. The flesh, too, is sweeter than that proceeding from any other food, with the exception of that which proceeds from buck wheat, a grain little used in England. That proceeding from Indian corn is, indeed, still sweeter and finer; but this is wholly out of the question with us.

83. I am, by-and-by, to speak of the cow to be kept by the labourer in husbandry. Then there will be milk to wet the bread with, an exceedingly great improvement in its taste as well as in its quality! This, of all the ways of using skim milk, is the most advantageous: and this great advantage must be wholly thrown away, if the bread of the family be bought at the shop. With milk, bread with very little wheat in it may be made far better than baker's bread; and, leaving the milk out of the question, taking a third of each sort of grain, you would get bread weighing as much as fourteen quartern loaves, for about 5s. 9d. at present prices of grain; that is to say, you would get it for about 5d. the quartern loaf, all expenses included; thus you have nine pounds and ten ounces of bread a day for about 5s. 9d. a week. Here is enough for a very large family. Very few labourers' families can want so much as this, unless indeed there be several persons in it capable of earning something by their daily labour. Here is cut and come again. Here is bread always for the table. Bread to carry a field; always a hunch of bread ready to put into the hand of a hungry child. We hear a great deal about "children crying for bread," and objects of compassion they and their parents are, when the latter have not the means of obtaining a sufficiency of bread. But I should be glad to be informed, how it is possible for a labouring man, who earns, upon an average, 10s. a week, who has not more than four children (and if he have more, some ought to be doing something;) who has a garden of a quarter of an acre of land (for that makes part of my plan;) who has a wife as industrious as she ought to be; who does not waste his earnings at the ale-house or the tea shop: I should be glad to know how such a man, while wheat shall be at the price of about 6s. a bushel, can possibly have children crying for bread!

84. Cry, indeed, they must, if he will persist in giving 13s. for a bushel of bread instead of 5s. 9d. Such a man is not to say that the bread which I have described is not good enough. It was good enough for his forefathers, who were too proud to be paupers, that is to say, abject and willing slaves. "Hogs eat barley." And hogs will eat wheat, too, when they can get at it. Convicts in condemned cells eat wheaten bread; but we think it no degradation to eat wheaten bread, too. I am for depriving the labourer of none of his rights; I would have him oppressed in no manner or shape; I would have him bold and free; but to have him such, he must have bread in his house, sufficient for all his family, and whether that bread be fine or coarse must depend upon the different circumstances which present themselves in the cases of different individuals.

85. The married man has no right to expect the same plenty of food and of raiment that the single man has. The time before marriage is the time to lay by, or, if the party choose, to indulge himself in the absence of labour. To marry is a voluntary act, and it is attended in the result with great pleasures and advantages. If, therefore, the laws be fair and equal; if the state of things be such that a labouring man can, with the usual ability of labourers, and with constant

industry, care and sobriety; with decency of deportment towards all his neighbours, cheerful obedience to his employer, and a due subordination to the laws; if the state of things be such, that such a man's earnings be sufficient to maintain himself and family with food, raiment, and lodging needful for them; such a man has no reason to complain; and no labouring man has reason to complain, if the numerousness of his family should call upon him for extraordinary exertion, or for frugality uncommonly rigid. The man with a large family has, if it be not in a great measure his own fault, a greater number of pleasures and of blessings than other men. If he be wise, and just as well as wise, he will see that it is reasonable for him to expect less delicate fare than his neighbours, who have a less number of children, or no children at all. He will see the justice as well as the necessity of his resorting to the use of coarser bread, and thus endeavour to make up that, or at least a part of that, which he loses in comparison with his neighbours. The quality of the bread ought, in every case, to be proportioned to the number of the family and the means of the head of that family. Here is no injury to health proposed; but, on the contrary, the best security for its preservation. Without bread, all is misery. The Scripture truly calls it the staff of life; and it may be called, too, the pledge of peace and happiness in the labourer's dwelling.

86. As to the act of making bread, it would be shocking indeed if that had to be taught by the means of books. Every woman, high or low, ought to know how to make bread. If she do not, she is unworthy of trust and confidence; and, indeed, a mere burden upon the community. Yet, it is but too true, that many women, even amongst those who have to get their living by their labour, know nothing of the making of bread; and seem to understand little more about it than the part which belongs to its consumption. A Frenchman, a Mr. CUSAR, who had been born in the West Indies, told me, that till he came to Long Island, he never knew how the flour came: that he was surprised when he learnt that it was squeezed out of little grains that grew at the tops of straw; for that he had always had an idea that it was got out of some large substances, like the yams that grow in tropical climates. He was a very sincere and good man, and I am sure he told me truth. And this may be the more readily believed, when we see so many women in England, who seem to know no more of the constituent parts of a loaf than they know of those of the moon. Servant women in abundance appear to think that loaves are made by the baker, as knights are made by the king; things of their pure creation, a creation, too, in which no one else can participate. Now, is not this an enormous evil? And whence does it come? Servant women are the children of the labouring classes; and they would all know how to make bread, and know well how to make it too, if they had been fed on bread of their mother's and their own making.

87. How serious a matter, then, is this, even in this point of view! A servant that cannot make bread is not entitled to the same wages as one that can. If she can neither bake nor brew; if she be ignorant of the nature of flour, yeast, malt, and hops, what is she good for? If she understand these matters well; if she be able to supply her employer with bread and with beer, she is really valuable; she is entitled to good wages, and to consideration and respect into the bargain; but if she be wholly deficient in these particulars, and can merely dawdle about with a bucket and a broom, she can be of very little consequence; to lose her, is merely to lose a consumer of food, and she can expect very little indeed in the way of desire to make her life easy and pleasant. Why should any one have such desire? She is not a child of the family. She is not a relation. Any one as well as she can take in a loaf from the baker, or a barrel of beer from the brewer. She has nothing whereby to bind her employer to her. To sweep a room any thing is

capable of that has got two hands. In short, she has no useful skill, no useful ability; she is an ordinary drudge, and she is treated accordingly.

88. But, if such be her state in the house of an employer, what is her state in the house of a husband? The lover is blind; but the husband has eyes to see with. He soon discovers that there is something wanted besides dimples and cherry cheeks; and I would have fathers seriously reflect, and to be well assured, that the way to make their daughters to be long admired, beloved and respected by their husbands, is to make them skilful, able and active in the most necessary concerns of a family. Eating and drinking come three times every day; the preparations for these, and all the ministry necessary to them, belong to the wife; and I hold it to be impossible, that at the end of two years, a really ignorant, sluttish wife should possess any thing worthy of the name of love from her husband. This, therefore, is a matter of far greater moment to the father of a family, than, whether the Parson of the parish, or the Methodist Priest, be the most "Evangelical" of the two; for it is here a question of the daughter's happiness or misery for life. And I have no hesitation to say, that if I were a labouring man, I should prefer teaching my daughters to bake, brew, milk, make butter and cheese, to teaching them to read the Bible till they had got every word of it by heart; and I should think, too, nay I should know, that I was in the former case doing my duty towards God as well as towards my children.

89. When we see a family of dirty, ragged little creatures, let us inquire into the cause; and ninety-nine times out of every hundred we shall find that the parents themselves have been brought up in the same way. But a consideration which ought of itself to be sufficient, is the contempt in which a husband will naturally hold a wife that is ignorant of the matters necessary to the conducting of a family. A woman who understands all the things above mentioned, is really a skilful person; a person worthy of respect, and that will be treated with respect too, by all but brutish employers or brutish husbands; and such, though sometimes, are not very frequently found. Besides, if natural justice and our own interest had not the weight which they have, such valuable persons will be treated with respect. They know their own worth; and, accordingly, they are more careful of their character, more careful not to lessen by misconduct the value which they possess from their skill and ability.

90. Thus, then, the interest of the labourer; his health; the health of his family; the peace and happiness of his home; the prospects of his children through life; their skill, their ability, their habits of cleanliness, and even their moral deportment; all combine to press upon him the adoption and the constant practice of this branch of domestic economy. "Can she bake?" is the question that I always put. If she can, she is worth a pound or two a year more. Is that nothing? Is it nothing for a labouring man to make his four or five daughters worth eight or ten pounds a year more; and that too while he is by the same means providing the more plentifully for himself and the rest of his family? The reasons on the side of the thing that I contend for are endless; but if this one motive be not sufficient, I am sure, all that I have said, and all that I could say, must be wholly unavailing.

91. Before, however, I dismiss this subject, let me say a word or two to those persons who do not come under the denomination of labourers. In London, or in any very large town where the space is so confined, and where the proper fuel is not handily to be come at and stored for use, to bake your own bread may be attended with too much difficulty; but in all other situations there appears to me to be hardly any excuse for not baking bread at home. If the family

consist of twelve or fourteen persons, the money actually saved in this way (even at present prices) would be little short of from twenty to thirty pounds a year. At the utmost here is only the time of one woman occupied one day in the week. Now mind, here are twenty-five pounds to be employed in some way different from that of giving it to the baker. If you add five of these pounds to a woman's wages, is not that full as well employed as giving it in wages to the baker's men? Is it not better employed for you? and is it not better employed for the community? It is very certain, that if the practice were as prevalent as I could wish, there would be a large deduction from the regular baking population; but would there be any harm if less alum were imported into England, and if some of those youths were left at the plough, who are now bound in apprenticeships to learn the art and mystery of doing that which every girl in the kingdom ought to be taught to do by her mother? It ought to be a maxim with every master and every mistress, never to employ another to do that which can be done as well by their own servants. The more of their money that is retained in the hands of their own people, the better it is for them altogether. Besides, a man of a right mind must be pleased with the reflection, that there is a great mass of skill and ability under his own roof. He feels stronger and more independent on this account, all pecuniary advantage out of the question. It is impossible to conceive any thing more contemptible than a crowd of men and women living together in a house, and constantly looking out of it for people to bring them food and drink, and to fetch their garments to and fro. Such a crowd resemble a nest of unfledged birds, absolutely dependent for their very existence on the activity and success of the old ones.

92. Yet, on men go, from year to year, in this state of wretched dependence, even when they have all the means of living within themselves, which is certainly the happiest state of life that any one can enjoy. It may be asked, Where is the mill to be found? where is the wheat to be got? The answer is, Where is there not a mill? where is there not a market? They are every where, and the difficulty is to discover what can be the particular attractions contained in that long and luminous manuscript, a baker's half-yearly bill.

93. With regard to the mill, in speaking of families of any considerable number of persons, the mill has, with me, been more than once a subject of observation in print. I for a good while experienced the great inconvenience and expense of sending my wheat and other grain to be ground at a mill. This expense, in case of a considerable family, living at only a mile from a mill, is something; but the inconveniency and uncertainty are great. In my "Year's Residence in America," from Paragraphs 1031 and onwards, I give an account of a horse-mill which I had in my farm yard; and I showed, I think very clearly, that corn could be ground cheaper in this way than by wind or water, and that it would answer well to grind for sale in this way as well as for home use. Since my return to England I have seen a mill, erected in consequence of what the owner had read in my book. This mill belongs to a small farmer, who, when he cannot work on his land with his horses, or in the season when he has little for them to do, grinds wheat, sells the flour; and he takes in grists to grind, as other millers do. This mill goes with three small horses; but what I would recommend to gentlemen with considerable families, or to farmers, is a mill such as I myself have at present.

94. With this mill, turned by a man and a stout boy, I can grind six bushels of wheat in a day and dress the flour. The grinding of six bushels of wheat at ninepence a bushel comes to four and sixpence, which pays the man and the boy, supposing them (which is not and seldom

can be the case) to be hired for the express purpose out of the street. With the same mill you grind meat for your pigs; and of this you will get eight or ten bushels ground in a day. You have no trouble about sending to the mill; you are sure to have your own wheat; for strange as it may seem, I used sometimes to find that I sent white Essex wheat to the mill, and that it brought me flour from very coarse red wheat. There is no accounting for this, except by supposing that wind and water power has something in it to change the very nature of the grain; as, when I came to grind by horses, such as the wheat went into the hopper, so the flour came out into the bin.

95. But mine now is only on the petty scale of providing for a dozen of persons and a small lot of pigs. For a farm-house, or a gentleman's house in the country, where there would be room to have a walk for a horse, you might take the labour from the men, clap any little horse, pony, or even ass to the wheel; and he would grind you off eight or ten bushels of wheat in a day, and both he and you would have the thanks of your men into the bargain.

96. The cost of this mill is twenty pounds. The dresser is four more; the horse-path and wheel might, possibly, be four or five more; and, I am very certain, that to any farmer living at a mile from a mill, (and that is less than the average distance perhaps;) having twelve persons in family, having forty pigs to feed, and twenty hogs to fatten, the savings of such a mill would pay the whole expenses of it the very first year. Such a farmer cannot send less than fifty times a year to the mill. Think of that, in the first place! The elements are not always propitious: sometimes the water fails, and sometimes the wind. Many a farmer's wife has been tempted to vent her spleen on both. At best, there must be horse and man, or boy, and, perhaps, cart, to go to the mill; and that, too, observe, in all weathers, and in the harvest as well as at other times of the year. The case is one of imperious necessity: neither floods nor droughts, nor storms nor calms, will allay the cravings of the kitchen, nor quiet the clamorous uproar of the stye. Go, somebody must, to some place or other, and back they must come with flour and with meal. One summer many persons came down the country more than fifty miles to a mill that I knew in Pennsylvania; and I have known farmers in England carry their grists more than fifteen miles to be ground. It is surprising, that, under these circumstances, hand-mills and horse-mills should not, long ago, have become of more general use; especially when one considers that the labour, in this case, would cost the farmer next to nothing. To grind would be the work of a wet day. There is no farmer who does not at least fifty days in every year exclaim, when he gets up in the morning, "What shall I set them at to-day?" If he had a mill, he would make them pull off their shoes, sweep all out clean, winnow up some corn, if he had it not already done, and grind and dress, and have every thing in order. No scolding within doors about the grist; no squeaking in the stye; no boy sent off in the rain to the mill.

97. But there is one advantage which I have not yet mentioned; and which is the greatest of all; namely, that you would have the power of supplying your married labourers; your blacksmith's men sometimes; your wheelwright's men at other times; and, indeed, the greater part of the persons that you employed, with good flour, instead of their going to purchase their flour, after it had passed through the hands of a Corn Merchant, a Miller, a Flour Merchant, and a Huckster, every one of whom does and must have a profit out of the flour, arising from wheat grown upon, and sent away from, your very farm! I used to let all my people have flour at the same price that they would otherwise have been compelled to give for worse flour. Every Farmer

will understand me when I say, that he ought to pay for nothing in money, which he can pay for in any thing but money. His maxim is to keep the money that he takes as long as he can.

Now here is a most effectual way of putting that maxim in practice to a very great extent. Farmers know well that it is the Saturday night which empties their pockets; and here is the means of cutting off a good half of the Saturday night. The men have better flour for the same money, and still the farmer keeps at home those profits which would go to the maintaining of the dealers in wheat and in flour.

98. The maker of my little mill is Mr. HILL, of Oxford-street. The expense is what I have stated it to be. I, with my small establishment, find the thing convenient and advantageous; what then must it be to a gentleman in the country who has room and horses, and a considerable family to provide for? The dresser is so contrived as to give you at once, meal, of four degrees of fineness; so that, for certain purposes, you may take the very finest; and, indeed, you may have your flour, and your bread of course, of what degree of fineness you please. But there is also a steel mill, much less expensive, requiring less labour, and yet quite sufficient for a family. Mills of this sort, very good, and at a reasonable price, are to be had of Mr. PARKES, in Fenchurch-street, London. These are very complete things of their kind. Mr. PARKES has, also, excellent Malt-Mills.

99. In concluding this part of my Treatise, I cannot help expressing my hope of being instrumental in inducing a part of the labourers, at any rate, to bake their own bread; and, above all things, to abandon the use of "Ireland's lazy root." Nevertheless, so extensive is the erroneous opinion relative to this villanous root, that I really began to despair of checking its cultivation and use, till I saw the declaration which Mr. WAKEFIELD had the good sense and the spirit to make before the "AGRICULTURAL COMMITTEE." Be it observed, too, that Mr. WAKEFIELD had himself made a survey of the state of Ireland. What he saw there did not encourage him, doubtless, to be an advocate for the growing of this root of wretchedness. It is an undeniable fact, that, in the proportion that this root is in use, as a substitute for bread, the people are wretched; the reasons for which I have explained and enforced a hundred times over. Mr. WILLIAM HANNING told the Committee that the labourers in his part of Somersetshire were "almost wholly supplied with potatoes, breakfast and dinner, brought them in the fields, and nothing but potatoes; and that they used, in better times, to get a certain portion of bacon and cheese, which, on account of their "poverty, they do not eat now." It is impossible that men can be contented in such a state of things: it is unjust to desire them to be contented: it is a state of misery and degradation to which no part of any community can have any show of right to reduce another part: men so degraded have no protection; and it is a disgrace to form part of a community to which they belong. This degradation has been occasioned by a silent change in the value of the money of the country. This has purloined the wages of the labourer; it has reduced him by degrees to housel with the spider and the bat, and to feed with the pig. It has changed the habits, and, in a great measure, the character of the people. The sins of this system are enormous and undescribable; but, thank God! they seem to be approaching to their end! Money is resuming its value, labour is recovering its price: let us hope that the wretched potatoe is disappearing, and that we shall, once more, see the knife in the labourer's hand and the loaf upon his board.

[This was written in 1821. Now (1823) we have had the experience of 1822, when, for the first time, the world saw a considerable part of a people, plunged into all the horrors of famine, at a moment when the government of that nation declared food to be abundant! Yes, the year 1822 saw Ireland in this state; saw the people of whole parishes receiving the extreme unction preparatory to yielding up their breath for want of food; and this while large exports of meat and flour were taking place in that country! But horrible as this was, disgraceful as it was to the name of Ireland, it was attended with this good effect: it brought out, from many members of Parliament (in their places,) and from the public in general, the acknowledgment, that the misery and degradation of the Irish were chiefly owing to the use of the potatoe as the almost sole food of the people.]

100. In my next number I shall treat of the keeping of cows. I have said that I will teach the cottager how to keep a cow all the year round upon the produce of a quarter of an acre, or, in other words, forty rods, of land; and, in my next, I will make good my promise.

NO. IV

MAKING BREAD--(CONTINUED.)

101. In the last number, at Paragraph 86, I observed that I hoped it was unnecessary for me to give any directions as to the mere act of making bread. But several correspondents inform me that, without these directions, a conviction of the utility of baking bread at home is of no use to them. Therefore, I shall here give those directions, receiving my instructions here from one, who, I thank God, does know how to perform this act.

102. Suppose the quantity be a bushel of flour. Put this flour into a trough that people have for the purpose, or it may be in a clean smooth tub of any shape, if not too deep, and if sufficiently large. Make a pretty deep hole in the middle of this heap of flour. Take (for a bushel) a pint of good fresh yeast, mix it and stir it well up in a pint of soft water milk-warm. Pour this into the hole in the heap of flour. Then take a spoon and work it round the outside of this body of moisture so as to bring into that body, by degrees, flour enough to make it form a thin batter, which you must stir about well for a minute or two. Then take a handful of flour and scatter it thinly over the head of this batter, so as to hide it. Then cover the whole over with a cloth to keep it warm; and this covering, as well as the situation of the trough, as to distance from the fire, must depend on the nature of the place and state of the weather as to heat and cold. When you perceive that the batter has risen enough to make cracks in the flour that you covered it over with, you begin to form the whole mass into dough, thus: you begin round the hole containing the batter, working the flour into the batter, and pouring in, as it is wanted to make the flour mix with the batter, soft water milk-warm, or milk, as hereafter to be mentioned. Before you begin this, you scatter the salt over the heap at the rate of half a pound to a bushel of flour. When you have got the whole sufficiently moist, you knead it well.

This is a grand part of the business; for, unless the dough be well worked, there will be little round lumps of flour in the loaves; and, besides, the original batter, which is to give fermentation to the whole, will not be duly mixed. The dough must, therefore, be well worked. The fists must go heartily into it. It must be rolled over; pressed out; folded up and pressed out again, until it be completely mixed, and formed into a stiff and tough dough. This is labour, mind. I have never quite liked baker's bread since I saw a great heavy fellow, in a bakehouse in France, kneading bread with his naked feet! His feet looked very white, to be sure: whether they were of that colour before he got into the trough I could not tell. God forbid, that I should suspect that this is ever done in England! It is labour; but, what is exercise other than labour? Let a young woman bake a bushel once a week, and she will do very well without phials and gallipots.

103. Thus, then, the dough is made. And, when made, it is to be formed into a lump in the middle of the trough, and, with a little dry flour thinly scattered over it, covered over again to be kept warm and to ferment; and in this state, if all be done rightly, it will not have to remain more than about 15 or 20 minutes.

104. In the mean while the oven is to be heated; and this is much more than half the art of the operation. When an oven is properly heated, can be known only by actual observation. Women who understand the matter, know when the heat is right the moment they put their faces within a yard of the oven-mouth; and once or twice observing is enough for any person of common capacity. But this much may be said in the way of rule: that the fuel (I am supposing a brick oven) should be dry (not rotten) wood, and not mere brush-wood, but rather fagot-sticks. If larger wood, it ought to be split up into sticks not more than two, or two and a half inches through.

Bush-wood that is strong, not green and not too old, if it be hard in its nature and has some sticks in it, may do. The woody parts of furze, or ling, will heat an oven very well. But the thing is, to have a lively and yet somewhat strong fire; so that the oven may be heated in about 15 minutes, and retain its heat sufficiently long.

105. The oven should be hot by the time that the dough, as mentioned in Paragraph 103, has remained in the lump about 20 minutes. When both are ready, take out the fire, and wipe the oven out clean, and, at nearly about the same moment, take the dough out upon the lid of the baking trough, or some proper place, cut it up into pieces, and make it up into loaves, kneading it again into these separate parcels; and, as you go on, shaking a little flour over your board, to prevent the dough from adhering to it. The loaves should be put into the oven as quickly as possible after they are formed; when in, the oven-lid, or door, should be fastened up very closely; and, if all be properly managed, loaves of about the size of quartern loaves will be sufficiently baked in about two hours. But they usually take down the lid, and look at the bread, in order to see how it is going on.

106. And what is there worthy of the name of plague, or trouble, in all this? Here is no dirt, no filth, no rubbish, no litter, no slop. And, pray, what can be pleasanter to behold? Talk, indeed, of your pantomimes and gaudy shows; your processions and installations and coronations! Give me, for a beautiful sight, a neat and smart woman, heating her oven and setting in her bread! And, if the bustle does make the sign of labour glisten on her brow, where is the man that would not kiss that off, rather than lick the plaster from the cheek of a duchess.

107. And what is the result? Why, good, wholesome food, sufficient for a considerable family for a week, prepared in three or four hours. To get this quantity of food, fit to be eaten, in the shape of potatoes, how many fires! what a washing, what a boiling, what a peeling, what a slopping, and what a messing! The cottage everlastingly in a litter; the woman's hands everlastingly wet and dirty; the children grimed up to the eyes with dust fixed on by potato-starch; and ragged as colts, the poor mother's time all being devoted to the everlasting boilings of the pot! Can any man, who knows any thing of the labourer's life, deny this? And will, then, any body, except the old shuffle-breeches band of the Quarterly Review, who have all their lives been moving from garret to garret, who have seldom seen the sun, and never the dew except in

print; will any body except these men say, that the people ought to be taught to use potatoes as a substitute for bread?

BREWING BEER.

108. This matter has been fully treated of in the two last numbers. But several correspondents wishing to fall upon some means of rendering the practice beneficial to those who are unable to purchase brewing utensils, have recommended the lending of them, or letting out, round a neighbourhood. Another correspondent has, therefore, pointed out to me an Act of Parliament which touches upon this subject; and, indeed, what of Excise Laws and Custom Laws and Combination Laws and Libel Laws, a human being in this country scarcely knows what he dares do or what he dares say. What father, for instance, would have imagined, that, having brewing utensils, which two men carry from house to house as easily as they can a basket, he dared not lend them to his son, living in the next street, or at the next door? Yet such really is the law; for, according to the Act 5th of the 22 and 23 of that honest and sincere gentleman Charles II., there is a penalty of 50l. for lending or letting brewing utensils. However, it has this limit; that the penalty is confined to Cities, Corporate Towns, and Market Towns, WHERE THERE IS A PUBLIC BREWHOUSE. So that, in the first place, you may let, or lend, in any place where there is no public brewhouse; and in all towns not corporate or market, and in all villages, hamlets, and scattered places.

109. Another thing is, can a man who has brewed beer at his own house in the country, bring that beer into town to his own house, and for the use of his family there? This has been asked of me. I cannot give a positive answer without reading about seven large volumes in quarto of taxing laws. The best way would be to try it; and, if any penalty, pay it by subscription, if that would not come under the law of conspiracy! However, I think, there can be no danger here. So monstrous a thing as this can, surely, not exist. If there be such a law, it is daily violated; for nothing is more common than for country gentlemen, who have a dislike to die by poison, bringing their home-brewed beer to London.

110. Another correspondent recommends parishes to make their own malt. But, surely, the landlords mean to get rid of the malt and salt tax! Many dairies, I dare say, pay 50l. a year each in salt tax. How, then, are they to contend against Irish butter and Dutch butter and cheese? And as to the malt tax, it is a dreadful drain from the land. I have heard of labourers, living "in unkent places," making their own malt, even now! Nothing is so easy as to make your own malt, if you were permitted. You soak the barley about three days (according to the state of the weather.) and then you put it upon stones or bricks and keep it turned, till the root shoots out; and then to know when to stop, and to put it to dry, take up a corn (which you will find nearly transparent) and look through the skin of it. You will see the spear, that is to say, the shoot that would come out of the ground, pushing on towards the point of the barley-corn. It starts from the bottom, where the root comes out; and it goes on towards the other end; and would, if kept moist, come out at that other end when the root was about an inch long. So that, when you have got the root to start, by soaking and turning in heap, the spear is on its way. If you look in through the skin, you will see it; and now observe; when the point of the spear has got along as far as the middle of the barley-corn, you should take your barley and dry it. How easy would every family, and especially every farmer, do this, if it were not for the punishment attached to it! The persons in the "unkent places" before mentioned, dry the malt in their oven! But let us hope that the

labourer will soon be able to get malt without exposing himself to punishment as a violater of the law.

KEEPING COWS.

111. As to the use of milk and of that which proceeds from milk, in a family, very little need be said. At a certain age bread and milk are all that a child wants. At a later age they furnish one meal a day for children. Milk is, at all seasons, good to drink. In the making of puddings, and in the making of bread too, how useful is it! Let any one who has eaten none but baker's bread for a good while, taste bread home-baked, mixed with milk instead of with water; and he will find what the difference is. There is this only to be observed, that in hot weather, bread mixed with milk will not keep so long as that mixed with water. It will of course turn sour sooner.

112. Whether the milk of a cow be to be consumed by a cottage family in the shape of milk, or whether it be to be made to yield butter, skim-milk, and buttermilk, must depend on circumstances. A woman that has no child, or only one, would, perhaps, find it best to make some butter at any rate.

Besides, skim-milk and bread (the milk being boiled) is quite strong food enough for any children's breakfast, even when they begin to go to work; a fact which I state upon the most ample and satisfactory experience, very seldom having ever had any other sort of breakfast myself till I was more than ten years old, and I was in the fields at work full four years before that. I will here mention that it gave me singular pleasure to see a boy, just turned of six, helping his father to reap, in Sussex, this last summer. He did little, to be sure; but it was something. His father set him into the ridge at a great distance before him; and when he came up to the place, he found a sheaf cut; and, those who know what it is to reap, know how pleasant it is to find now and then a sheaf cut ready to their hand. It was no small thing to see a boy fit to be trusted with so dangerous a thing as a reap-hook in his hands, at an age when "young masters" have nursery-maids to cut their victuals for them, and to see that they do not fall out of the window, tumble down stairs, or run under carriage-wheels or horses' bellies. Was not this father discharging his duty by this boy much better than he would have been by sending him to a place called a school? The boy is in a school here; and an excellent school too: the school of useful labour. I must hear a great deal more than I ever have heard, to convince me, that teaching children to read tends so much to their happiness, their independence of spirit, their manliness of character, as teaching them to reap. The creature that is in want must be a slave; and to be habituated to labour cheerfully is the only means of preventing nineteen-twentieths of mankind from being in want. I have digressed here; but observations of this sort can, in my opinion, never be too often repeated; especially at a time when all sorts of mad projects are on foot, for what is falsely called educating the people, and when some would do this by a tax that would compel the single man to give part of his earnings to teach the married man's children to read and write.

113. Before I quit the uses to which milk may be put, let me mention, that, as mere drink, it is, unless perhaps in case of heavy labour, better, in my opinion, than any beer, however good. I have drinked little else for the last five years, at any time of the day. Skim-milk I mean. If you have not milk enough to wet up your bread with (for a bushel of flour requires about 16 to 18 pints,) you make up the quantity with water, of course; or, which is a very good way, with

water that has been put, boiling hot, upon bran, and then drained off. This takes the goodness out of the bran to be sure; but really good bread is a thing of so much importance, that it always ought to be the very first object in domestic economy.

114. The cases vary so much, that it is impossible to lay down rules for the application of the produce of a cow, which rules shall fit all cases. I content myself, therefore, with what has already been said on this subject; and shall only make an observation on the act of milking, before I come to the chief matter; namely, the getting of the food for the cow. A cow should be milked clean. Not a drop, if it can be avoided, should be left in the udder. It has been proved that the half pint that comes out last has twelve times, I think it is, as much butter in it, as the half pint that comes out first. I tried the milk of ten Alderney cows, and, as nearly as I, without being very nice about the matter, could ascertain, I found the difference to be about what I have stated. The udder would seem to be a sort of milk-pan in which the cream is uppermost, and, of course, comes out last, seeing that the outlet is at the bottom. But, besides this, if you do not milk clean, the cow will give less and less milk, and will become dry much sooner than she ought. The cause of this I do not know, but experience has long established the fact.

115. In providing food for a cow we must look, first, at the sort of cow; seeing that a cow of one sort will certainly require more than twice as much food as a cow of another sort. For a cottage, a cow of the smallest sort common in England is, on every account, the best; and such a cow will not require above 70 or 80 pounds of good moist food in the twenty-four hours.

116. Now, how to raise this food on 40 rods of ground is what we want to know. It frequently happens that a labourer has more than 40 rods of ground. It more frequently happens, that he has some common, some lane, some little out-let or other, for a part of the year, at least. In such cases he may make a different disposition of his ground; or may do with less than the 40 rods. I am here, for simplicity's sake, to suppose, that he have 40 rods of clear, unshaded land, besides what his house and sheds stand upon; and that he have nothing further in the way of means to keep his cow.

117. I suppose the 40 rods to be clean and unshaded; for I am to suppose, that when a man thinks of 5 quarts of milk a day, on an average, all the year round, he will not suffer his ground to be encumbered by apple-trees that give him only the means of treating his children to fits of the belly-ache, or with currant and gooseberry bushes, which, though their fruit do very well to amuse, really give nothing worthy of the name of food, except to the blackbirds and thrushes. The ground is to be clear of trees; and, in the spring, we will suppose it to be clean. Then, dig it up deeply, or, which is better, trench it, keeping, however, the top spit of the soil at the top. Lay it in ridges in April or May about two feet apart, and made high and sharp. When the weeds appear about three inches high, turn the ridges into the furrows (never moving the ground but in dry weather,) and bury all the weeds. Do this as often as the weeds get three inches high; and by the fall, you will have really clean ground, and not poor ground.

118. There is the ground then, ready. About the 26th of August, but not earlier, prepare a rod of your ground; and put some manure in it (for some you must have,) and sow one half of it with Early York Cabbage Seed, and the other half with Sugar-loaf Cabbage Seed, both of the true sort, in little drills at 8 inches apart, and the seeds thin in the drill. If the plants come up at two inches apart (and they should be thinned if thicker,) you will have a plenty. As soon as fairly

out of ground, hoe the ground nicely, and pretty deeply, and again in a few days. When the plants have six leaves, which will be very soon, dig up, make fine, and manure another rod or two, and prick out the plants, 4000 of each in rows at eight inches apart and three inches in the row. Hoe the ground between them often, and they will grow fast and be straight and strong. I suppose that these beds for plants take 4 rods of your ground. Early in November, or, as the weather may serve, a little earlier or later, lay some manure (of which I shall say more hereafter) between the ridges, in the other 36 rods, and turn the ridges over on this manure, and then transplant your plants on the ridges at 15 inches apart. Here they will stand the winter; and you must see that the slugs do not eat them. If any plants fail, you have plenty in the bed where you prick them out; for your 36 rods will not require more than 4000 plants. If the winter be very hard, and bad for plants, you cannot cover 36 rods; but you may the bed where the rest of your plants are. A little litter, or straw, or dead grass, or fern, laid along between the rows and the plants, not to cover the leaves, will preserve them completely. When people complain of all their plants being "cut off," they have, in fact nothing to complain of but their own extreme carelessness. If I had a gardener who complained of all his plants being cut off, I should cut him off pretty quickly. If those in the 36 rods fail, or fail in part, fill up their places, later in the winter, by plants from the bed.

119. If you find the ground dry at the top during the winter, hoe it, and particularly near the plants, and rout out all slugs and insects. And when March comes, and the ground is dry, hoe deep and well, and earth the plants up close to the lower leaves. As soon as the plants begin to grow, dig the ground with a spade clean and well, and let the spade go as near to the plants as you can without actually displacing the plants. Give them another digging in a month; and, if weeds come in the mean-while, hoe, and let not one live a week. Oh! "what a deal of work!" Well! but it is for yourself, and, besides, it is not all to be done in a day; and we shall by-and-by see what it is altogether.

120. By the first of June; I speak of the South of England, and there is also some difference in seasons and soils; but, generally speaking, by the first of June you will have turned-in cabbages, and soon you will have the Early Yorks solid. And by the first of June you may get your cow, one that is about to calve, or that has just calved, and at this time such a cow as you will want will not, thank God, cost above five pounds.

121. I shall speak of the place to keep her in, and of the manure and litter, by-and-by. At present I confine myself to her mere food. The 36 rods, if the cabbages all stood till they got solid, would give her food for 200 days, at 80 pounds weight per day, which is more than she would eat. But you must use some, at first, that are not solid; and, then, some of them will split before you can use them. But you will have pigs to help off with them, and to gnaw the heads of the stumps. Some of the sugar-loaves may have been planted out in the spring; and thus these 36 rods will get you along to some time in September.

122. Now mind, in March, and again in April, sow more Early Yorks, and get them to be fine stout plants, as you did those in the fall. Dig up the ground and manure it, and, as fast as you cut cabbages, plant cabbages; and in the same manner and with the same cultivation as before. Your last planting will be about the middle of August, with stout plants, and these will serve you into the month of November.

123. Now we have to provide from December to May inclusive; and that, too, out of this same piece of ground. In November there must be, arrived at perfection, 3000 turnip plants. These, without the greens, must weigh, on an average, 5 pounds, and this, at 80 pounds a day, will keep the cow 187 days; and there are but 182 days in these six months. The greens will have helped put the latest cabbages to carry you through November, and perhaps into December. But for these six months, you must depend on nothing but the Swedish turnips.

124. And now, how are these to be had upon the same ground that bears the cabbages? That we are now going to see. When you plant out your cabbages at the out-set, put first a row of Early Yorks, then a row of Sugar-loaves, and so on throughout the piece. Of course, as you are to use the Early Yorks first, you will cut every other row; and the Early Yorks that you are to plant in summer will go into the intervals. By-and-by the

Sugar-loaves are cut away, and in their place will come Swedish turnips, you digging and manuring the ground as in the case of the cabbages: and, at last, you will find about 16 rods where you will have found it too late, and unnecessary besides, to plant any second crop of cabbages. Here the Swedish turnips will stand in rows at two feet apart, (and always a foot apart in the row,) and thus you will have three thousand turnips; and if these do not weigh five pounds each on an average, the fault must be in the seed or in the management.

125. The Swedish turnips are raised in this manner. You will bear in mind the four rods of ground in which you have sowed and pricked out your cabbage plants. The plants that will be left there will, in April, serve you for greens, if you ever eat any, though bread and bacon are very good without greens, and rather better than with. At any rate, the pig, which has strong powers of digestion, will consume this herbage. In a part of these four rods you will, in March and April, as before directed, have sown and raised your Early Yorks for the summer planting. Now, in the last week of May, prepare a quarter of a rod of this ground, and sow it, precisely as directed for the Cabbage-seed, with Swedish turnip-seed; and sow a quarter of a rod every three days, till you have sowed two rods. If the fly appear, cover the rows over in the day-time with cabbage leaves, and take the leaves off at night; hoe well between the plants; and when they are safe from the fly, thin them to four inches apart in the row. The two rods will give you nearly five thousand plants, which is 2000 more than you will want. From this bed you draw your plants to transplant in the ground where the cabbages have stood, as before directed. You should transplant none much before the middle of July, and not much later than the middle of August. In the two rods, whence you take your turnip plants, you may leave plants to come to perfection, at two feet distances each way; and this will give you over and above, 840 pounds weight of turnips. For the other two rods will be ground enough for you to sow your cabbage plants in at the end of August, as directed for last year.

126. I should now proceed to speak of the manner of harvesting, preserving, and using the crops; of the manner of feeding the cow; of the shed for her; of the managing of the manure, and several other less important things; but these, for want of room here, must be reserved for the beginning of my next Number. After, therefore, observing that the Turnip plants must be transplanted in the same way that Cabbage plants are; and that both ought to be transplanted in dry weather and in ground just fresh digged, I shall close this Number with the notice of two points which I am most anxious to impress upon the mind of every reader.

127. The first is, whether these crops give an ill taste to milk and butter. It is very certain, that the taste and smell of certain sorts of cattle-food will do this; for, in some parts of America, where the wild garlick, of which the cows are very fond, and which, like other bulbous-rooted plants, springs before the grass, not only the milk and butter have a strong taste of garlick, but even the veal, when the calves suck milk from such sources. None can be more common expressions, than, in Philadelphia market, are those of

Garlicky Butter and Garlicky Veal, I have distinctly tasted the Whiskey in milk of cows fed on distiller's wash. It is also certain, that, if the cow eat putrid leaves of cabbages and turnips, the butter will be offensive. And the white-turnip, which is at best but a poor thing, and often half putrid, makes miserable butter. The large cattle-cabbage, which, when loaved hard, has a strong and even an offensive smell, will give a bad taste and smell to milk and butter, whether there be putrid leaves or not. If you boil one of these rank cabbages, the water is extremely offensive to the smell. But I state upon positive and recent experience, that Early York and Sugar-loaf Cabbages will yield as sweet milk and butter as any food that can be given to a cow. During this last summer, I have, with the exception about to be noticed, kept, from the 1st of May to the 22d of October, five cows upon the grass of two acres and a quarter of ground, the grass being generally cut up for them and given to them in the stall. I had in the spring 5000 cabbage plants, intended for my pigs, eleven in number. But the pigs could not eat half their allowance, though they were not very small when they began upon it. We were compelled to resort to the aid of the cows; and, in order to see the effect on the milk and butter, we did not mix the food; but gave the cows two distinct spells at the cabbages, each spell about 10 days in duration. The cabbages were cut off the stump with little or no care about dead leaves. And sweeter, finer butter, butter of a finer colour, than these cabbages made, never was made in this world. I never had better from cows feeding in the sweetest pasture. Now, as to Swedish turnips, they do give a little taste, especially if boiling of the milk pans be neglected, and if the greatest care be not taken about all the dairy tackle. Yet we have, for months together, had the butter so fine from Swedish turnips, that nobody could well distinguish it from grass-butter. But to secure this, there must be no sluttishness. Churn, pans, pail, shelves, wall, floor, and all about the dairy, must be clean; and, above all things, the pans must be boiled.

However, after all, it is not here a case of delicacy of smell so refined as to faint at any thing that meets it except the stink of perfumes. If the butter do taste a little of the Swedish turnip, it will do very well where there is plenty of that sweet sauce which early rising and bodily labour are ever sure to bring.

128. The other point (about which I am still more anxious) is the seed; for if the seed be not sound, and especially if it be not true to its kind, all your labour is in vain. It is best, if you can do it, to get your seed from some friend, or some one that you know and can trust. If you save seed, observe all the precautions mentioned in my book on Gardening. This very year I have some Swedish turnips, so called, about 7000 in number, and should, if my seed had been true, have had about twenty tons weight; instead of which I have about three! Indeed, they are not Swedish turnips, but a sort of mixture between that plant and rape. I am sure the seedsman did not wilfully deceive me. He was deceived himself. The truth is, that seedsmen are compelled to buy their seeds of this plant. Farmers save it; and they but too often pay very little attention to the manner of doing it. The best way is to get a dozen of fine turnip plants, perfect in all

respects, and plant them in a situation where the smell of the blossoms of nothing of the cabbage or rape or turnip or even charlock kind, can reach them. The seed will keep perfectly good for four years.

NO. V

KEEPING COWS--(CONTINUED.)

129. I have now, in the conclusion of this article, to speak of the manner of harvesting and preserving the Swedes; of the place to keep the cow in; of the manure for the land; and of the quantity of labour that the cultivation of the land and the harvesting of the crop will require.

130. Harvesting and preserving the Swedes. When they are ready to take up, the tops must be cut off, if not cut off before, and also the roots; but neither tops nor roots should be cut off very close. You will have room for ten bushels of the bulbs in the house, or shed. Put the rest into ten-bushel heaps. Make the heap upon the ground in a round form, and let it rise up to a point. Lay over it a little litter, straw, or dead grass, about three inches thick, and then earth upon that about six inches thick. Then cut a thin round green turf, about eighteen inches over, and put it upon the crown of the heap to prevent the earth from being washed off. Thus these heaps will remain till wanted for use. When given to the cow, it will be best to wash the Swedes and cut each into two or three pieces with a spade or some other tool. You can take in ten bushels at a time. If you find them sprouting in the spring, open the remaining heaps, and expose them to the sun and wind; and cover them again slightly with straw or litter of some sort.

131. As to the place to keep the cow in, much will depend upon situation and circumstances. I am always supposing that the cottage is a real cottage, and not a house in a town or village street; though, wherever there is the quarter of an acre of ground, the cow may be kept. Let me, however, suppose that which will generally happen; namely, that the cottage stands by the side of a road, or lane, and amongst fields and woods, if not on the side of a common. To pretend to tell a country labourer how to build a shed for a cow, how to stick it up against the end of his house, or to make it an independent erection; or to dwell on the materials, where poles, rods, wattles, rushes, furze, heath, and cooper-chips, are all to be gotten by him for nothing or next to nothing, would be useless; because a man who, thus situated, can be at any loss for a shed for his cow, is not only unfit to keep a cow, but unfit to keep a cat. The warmer the shed is the better it is. The floor should slope, but not too much. There are stones, of some sort or other, every-where, and about six wheel-barrow-fulls will pave the shed, a thing to be by no means neglected. A broad trough, or box, fixed up at the head of the cow, is the thing to give her food in; and she should be fed three times a day, at least; always at day-light and at sun-set. It is not absolutely necessary that a cow ever quit her shed, except just at calving time, or when taken to the bull. In the former case the time is, nine times out of ten, known to within forty-eight hours. Any enclosed field or place will do for her during a day or two; and for such purpose, if there be not room at home, no man will refuse place for her in a fallow field. It will, however, be good, where there is no common to turn her out upon, to have her led by a string, two or three times a week, which may be done by a child only five years old, to graze, or pick, along the

sides of roads and lanes. Where there is a common, she will, of course, be turned out in the day time, except in very wet or severe weather; and in a case like this, a smaller quantity of ground will suffice for the keeping of her. According to the present practice, a miserable "tallet" of bad hay is, in such cases, the winter provision for the cow. It can scarcely be called food; and the consequence is, the cow is both dry and lousy nearly half the year; instead of being dry only about fifteen days before calving, and being sleek and lusty at the end of the winter, to which a warm lodging greatly contributes. For, observe, if you keep a cow, any time between September and June, out in a field or yard, to endure the chances of the weather, she will not, though she have food precisely the same in quantity and quality, yield above two-thirds as much as if she were lodged in house; and in wet weather she will not yield half so much. It is not so much the cold as the wet that is injurious to all our stock in England.

132. The Manure. At the beginning this must be provided by collections made on the road; by the results of the residence in a cottage. Let any man clean out every place about his dwelling; rake and scrape and sweep all into a heap; and he will find that he has a great deal. Earth of almost any sort that has long lain on the surface, and has been trodden on, is a species of manure. Every act that tends to neatness round a dwelling, tends to the creating of a mass of manure. And I have very seldom seen a cottage, with a plat of ground of a quarter of an acre belonging to it, round about which I could not have collected a very large heap of manure. Every thing of animal or vegetable substance that comes into a house, must go out of it again, in one shape or another. The very emptying of vessels of various kinds, on a heap of common earth, makes it a heap of the best of manure. Thus goes on the work of reproduction; and thus is verified the words of the Scripture, "Flesh is grass, and there is nothing new under the sun." Thus far as to the outset. When you have got the cow, there is no more care about manure; for, and especially if you have a pig also, you must have enough annually for an acre of ground. And let it be observed, that, after a time, it will be unnecessary, and would be injurious, to manure for every crop; for that would produce more stalk and green than substantial part; as it is well known, that wheat plants, standing in ground too full of manure, will yield very thick and long straws, but grains of little or no substance. You ought to depend more on the spade and the hoe than on the dung-heap. Nevertheless, the greatest care should be taken to preserve the manure; because you will want straw, unless you be by the side of a common which gives you rushes, grassy furze, or fern; and to get straw you must give a part of your dung from the cow-stall and pig-sty. The best way to preserve manure, is to have a pit of sufficient dimensions close behind the cow-shed and pig-sty, for the run from these to go into, and from which all runs of rain water should be kept. Into this pit would go the emptying of the shed and of the sty, and the produce of all sweepings and cleanings round the house; and thus a large mass of manure would soon grow together. Much too large a quantity for a quarter of an acre of ground. One good load of wheat or rye straw is all that you would want for the winter, and half of one for the summer; and you would have more than enough dung to exchange against this straw.

133. Now, as to the quantity of labour that the cultivation of the land will demand in a year. We will suppose the whole to have five complete diggings, and say nothing about the little matters of sowing and planting and hoeing and harvesting, all which are a mere trifle. We are supposing the owner to be an able labouring man; and such a man will dig 12 rods of ground in a day. Here are 200 rods to be digged, and here are little less than 17 days of work at 12 hours in the day; or 200 hours' work, to be done in the course of the long days of spring and summer,

while it is light long before six in the morning, and long after six at night. What is it, then? Is it not better than time spent in the ale-house, or in creeping about after a miserable hare? Frequently, and most frequently, there will be a boy, if not two, big enough to help. And (I only give this as a hint) I saw, on the 7th of November last (1822,) a very pretty woman, in the village of Hannington, in Wiltshire, digging a piece of ground and planting it with Early Cabbages, which she did as handily and as neatly as any gardener that ever I saw. The ground was wet, and therefore, to avoid treading the digged ground in that state, she had her line extended, and put in the rows as she advanced in her digging, standing in the trench while she performed the act of planting, which she did with great nimbleness and precision. Nothing could be more skilfully or beautifully done. Her clothes were neat, clean, and tight about her. She had turned her handkerchief down from her neck, which, with the glow that the work had brought into her cheeks, formed an object which I do not say would have made me actually stop my chaise, had it not been for the occupation in which she was engaged; but, all taken together, the temptation was too strong to be resisted. But there is the Sunday; and I know of no law, human or divine, that forbids a labouring man to dig or plant his garden on Sunday, if the good of his family demand it; and if he cannot, without injury to that family, find other time to do it in. Shepherds, carters, pigfeeders, drovers, coachmen, cooks, footmen, printers, and numerous others, work on the Sundays. Theirs are deemed by the law works of necessity. Harvesting and haymaking are allowed to be carried on on the Sunday, in certain cases; when they are always carried on by provident farmers. And I should be glad to know the case which is more a case of necessity than that now under our view. In fact, the labouring people do work on the Sunday morning in particular, all over the country, at something or other, or they are engaged in pursuits a good deal less religious than that of digging and planting. So that, as to the 200 hours, they are easily found, without the loss of any of the time required for constant daily labour.

134. And what a produce is that of a cow! I suppose only an average of 5 quarts of milk a day. If made into butter, it will be equal every week to two days of the man's wages, besides the value of the skim milk: and this can hardly be of less value than another day's wages. What a thing, then, is this cow, if she earn half as much as the man! I am greatly under-rating her produce; but I wish to put all the advantages at the lowest. To be sure, there is work for the wife, or daughter, to milk and make butter. But the former is done at the two ends of the day, and the latter only about once in the week. And, whatever these may subtract from the labours of the field, which all country women ought to be engaged in whenever they conveniently can; whatever the cares created by the cow may subtract from these, is amply compensated for by the education that these cares will give to the children. They will all learn to milk,[7] and the girls to make butter. And which is a thing of the very first importance, they will all learn, from their infancy, to set a just value upon dumb animals, and will grow up in the habit of treating them with gentleness and feeding them with care. To those who have not been brought up in the midst of rural affairs, it is hardly possible to give an adequate idea of the importance of this part of education. I should be very loth to intrust the care of my horses, cattle, sheep, or pigs, to any one whose father never had cow or pig of his own. It is a general complaint, that servants, and especially farm-servants, are not so good as they used to be. How should they? They were formerly the sons and daughters of small farmers; they are now the progeny of miserable property-less labourers. They have never seen an animal in which they had any interest. They are careless by habit. This monstrous evil has arisen from causes which I have a thousand times

described; and which causes must now be speedily removed; or, they will produce a dissolution of society, and give us a beginning afresh.

135. The circumstances vary so much, that it is impossible to lay down precise rules suited to all cases. The cottage may be on the side of a forest or common; it may be on the side of a lane or of a great road, distant from town or village; it may be on the skirts of one of these latter: and then, again, the family may be few or great in number, the children small or big, according to all which circumstances, the extent and application of the cow-food, and also the application of the produce, will naturally be regulated. Under some circumstances, half the above crop may be enough; especially where good commons are at hand. Sometimes it may be the best way to sell the calf as soon as calved; at others, to fat it; and, at others, if you cannot sell it, which sometimes happens, to knock it on the head as soon as calved; for, where there is a family of small children, the price of a calf of two months old cannot be equal to the half of the value of the two months' milk. It is pure weakness to call it "a pity." It is a much greater pity to see hungry children crying for the milk that a calf is sucking to no useful purpose; and as to the cow and the calf, the one must lose her young, and the other its life, after all; and the respite only makes an addition to the sufferings of both.

136. As to the pretended unwholesomeness of milk in certain cases; as to its not being adapted to some constitutions, I do not believe one word of the matter. When we talk of the fruits, indeed, which were formerly the chief food of a great part of mankind, we should recollect, that those fruits grew in countries that had a sun to ripen the fruits, and to put nutritious matter into them. But as to milk, England yields to no country upon the face of the earth. Neat cattle will touch nothing that is not wholesome in its nature; nothing that is not wholly innoxious. Out of a pail that has ever had grease in it, they will not drink a drop, though they be raging with thirst. Their very breath is fragrance. And how, then, is it possible, that unwholesomeness should distil from the udder of a cow? The milk varies, indeed, in its quality and taste according to the variations in the nature of the food; but no food will a cow touch that is any way hostile to health.

Feed young puppies upon milk from the cow, and they will never die with that ravaging disease called "the distemper." In short, to suppose that milk contains any thing essentially unwholesome is monstrous. When, indeed, the appetite becomes vitiated: when the organs have been long accustomed to food of a more stimulating nature; when it has been resolved to eat ragouts at dinner, and drink wine, and to swallow "a devil," and a glass of strong grog at night; then milk for breakfast may be "heavy" and disgusting, and the feeder may stand in need of tea or laudanum, which differ only as to degrees of strength. But, and I speak from the most ample experience, milk is not "heavy," and much less is it unwholesome, when he who uses it rises early, never swallows strong drink, and never stuffs himself with flesh of any kind. Many and many a day I scarcely taste of meat, and then chiefly at breakfast, and that, too, at an early hour. Milk is the natural food of young people; if it be too rich, skim it again and again till it be not too rich. This is an evil easily cured. If you have now to begin with a family of children, they may not like it at first. But persevere; and the parent who does not do this, having the means in his hands, shamefully neglects his duty. A son who prefers a "devil" and a glass of grog to a hunch of bread and a bowl of cold milk, I regard as a pest; and for this pest the father has to thank himself.

137. Before I dismiss this article, let me offer an observation or two to those persons who live in the vicinity of towns, or in towns, and who, though they have large gardens, have

"no land to keep a cow," a circumstance which they "exceedingly regret." I have, I dare say, witnessed this case at least a thousand times. Now, how much garden ground does it require to supply even a large family with garden vegetables? The market gardeners round the metropolis of this wen-headed country; round this Wen of all wens;[8] round this prodigious and monstrous collection of human beings; these market gardeners have about three hundred thousand families to supply with vegetables, and these they supply well too, and with summer fruits into the bargain. Now, if it demanded ten rods to a family, the whole would demand, all but a fraction, nineteen thousand acres of garden ground. We have only to cast our eyes over what there is to know that there is not a fourth of that quantity. A square mile contains, leaving out parts of a hundred, 700 acres of land; and 19,000 acres occupy more than twenty-two square miles. Are there twenty-two square miles covered with the Wen's market gardens? The very question is absurd. The whole of the market gardens from Brompton to Hammersmith, extending to Battersea Rise on the one side, and to the Bayswater road on the other side, and leaving out loads, lanes, nurseries; pastures, corn-fields, and pleasure-grounds, do not, in my opinion, cover one square mile. To the north and south of the Wen there is very little in the way of market garden; and if, on both sides of the Thames, to the eastward of the Wen, there be three square miles actually covered with market gardens, that is the full extent. How, then, could the Wen be supplied, if it required ten rods to each family? To be sure, potatoes, carrots, and turnips, and especially the first of these, are brought, for the use of the Wen, from a great distance, in many cases. But, so they are for the use of the persons I am speaking of; for a gentleman thinks no more of raising a large quantity of these things in his garden, than he thinks of raising wheat there. How is it, then, that it requires half an acre, or eighty rods, in a private garden to supply a family, while these market gardeners supply all these families (and so amply too) from ten, or more likely, five rods of ground to a family? I have shown, in the last Number, that nearly fifteen tons of vegetables can be raised in a year upon forty rods of ground; that is to say, ten loads for a wagon and four good horses. And is not a fourth, or even an eighth, part of this weight, sufficient to go down the throats of a family in a year? Nay, allow that only a ton goes to a family in a year, it is more than six pound weight a day; and what sort of a family must that be that really swallows six pounds weight a day? and this a market gardener will raise for them upon less than three rods of ground; for he will raise, in the course of the year, even more than fifteen tons upon forty rods of ground. What is it, then, that they do with the eighty rods of ground in a private garden? Why, in the first place, they have one crop where they ought to have three. Then they do not half till the ground. Then they grow things that are not wanted. Plant cabbages and other things, let them stand till they be good for nothing, and then wheel them to the rubbish heap. Raise as many radishes, lettuces, and as much endive, and as many kidney-beans, as would serve for ten families; and finally throw nine-tenths of them away. I once saw not less than three rods of ground, in a garden of this sort, with lettuces all bearing seed. Seed enough for half a county. They cut a cabbage here and a cabbage there, and so let the whole of the piece of ground remain undug, till the last cabbage be cut. But, after all, the produce, even in this way, is so great, that it never could be gotten rid of, if the main part were not thrown away. The rubbish heap always receives four-fifths even of the eatable part of the produce.

138. It is not thus that the market gardeners proceed. Their rubbish heap consists of little besides mere cabbage stumps. No sooner is one crop on the ground than they settle in their minds what is to follow it. They clear as they go in taking off a crop, and, as they clear they dig and plant. The ground is never without seed in it or plants on it. And thus, in the course of the

year, they raise a prodigious bulk of vegetables from eighty rods of ground. Such vigilance and industry are not to be expected in a servant; for it is foolish to expect that a man will exert himself for another as much as he will for himself. But if I was situated as one of the persons is that I have spoken of in Paragraph 137; that is to say, if I had a garden of eighty rods, or even of sixty rods of ground, I would out of that garden, draw a sufficiency of vegetables for my family, and would make it yield enough for a cow besides. I should go a short way to work with my gardener. I should put Cottage Economy into his hands, and tell him, that if he could furnish me with vegetables, and my cow with food, he was my man; and that if he could not, I must get one that could and would. I am not for making a man toil like a slave; but what would become of the world, if a well-fed healthy man could exhaust himself in tilling and cropping and clearing half an acre of ground? I have known many men dig thirty rods of garden ground in a day; I have, before I was fourteen, digged twenty rods in a day, for more than ten days successively; and I have heard, and believe the fact, of a man at Portsea, who digged forty rods in one single day, between daylight and dark. So that it is no slavish toil that I am here recommending.

KEEPING PIGS.

139. Next after the Cow comes the Pig; and, in many cases, where a cow cannot be kept, a pig or pigs may be kept. But these are animals not to be ventured on without due consideration as to the means of feeding them; for a starved pig is a great deal worse than none at all. You cannot make bacon as you can milk, merely out of the garden. There must be something more. A couple of flitches of bacon are worth fifty thousand Methodist sermons and religious tracts. The sight of them upon the rack tends more to keep a man from poaching and stealing than whole volumes of penal statutes, though assisted by the terrors of the hulks and the gibbet. They are great softeners of the temper, and promoters of domestic harmony. They are a great blessing; but they are not to be had from herbage or roots of any kind; and, therefore, before a pig be attempted, the means ought to be considered.

140. Breeding sows are great favourites with Cottagers in general; but I have seldom known them to answer their purpose. Where there is an outlet, the sow will, indeed, keep herself by grazing in summer, with a little wash to help her out: and when her pigs come, they are many in number; but they are a heavy expense. The sow must live as well as a fatting hog, or the pigs will be good for little. It is a great mistake, too, to suppose that the condition of the sow previous to pigging is of no consequence; and, indeed, some suppose, that she ought to be rather bare of flesh at the pigging time. Never was a greater mistake; for if she be in this state, she presently becomes a mere rack of bones; and then, do what you will, the pigs will be poor things. However fat she may be before she farrow, the pigs will make her lean in a week. All her fat goes away in her milk, and unless the pigs have a store to draw upon, they pull her down directly; and, by the time they are three weeks old, they are starving for want; and then they never come to good.

141. Now, a cottager's sow cannot, without great expense, be kept in a way to enable her to meet the demands of her farrow. She may look pretty well; but the flesh she has upon her is not of the same nature as that which the farm-yard sow carries about her. It is the result of grass, and of poor grass, too, or other weak food; and not made partly out of corn and whey and strong wash, as in the case of the farmer's sow. No food short of that of a fatting hog will enable her to keep her pigs alive; and this she must have for ten weeks, and that at a great expense. Then comes the operation, upon the principle of Parson Malthus, in order to check population; and there is some risk here, though not very great. But there is the weaning; and who, that knows any thing about the matter, will think lightly of the weaning of a farrow of pigs! By having nice food given them, they seem, for a few days, not to miss their mother. But their appearance soon shows the want of her. Nothing but the very best food, and that given in the most judicious manner, will keep them up to any thing like good condition; and, indeed, there is nothing short of milk that will effect the thing well. How should it be otherwise? The very richest cow's milk is poor, compared with that of the sow; and, to be taken from this and put upon food, one ingredient of which is water, is quite sufficient to reduce the poor little things to bare bones and staring hair, a state to which cottagers' pigs very soon come in general; and, at last, he frequently drives them to market, and sells them for less than the cost of the food which they and the sow

have devoured since they were farrowed. It was, doubtless, pigs of this description that were sold the other day at Newbury market, for fifteen pence a piece, and which were, I dare say, dear even as a gift. To get such a pig to begin to grow will require three months, and with good feeding too in winter time. To be sure it does come to be a hog at last; but, do what you can, it is a dear hog.

142. The Cottager, then, can hold no competition with the Farmer in the breeding of pigs, to do which, with advantage, there must be milk, and milk, too, that can be advantageously applied to no other use. The cottager's pig must be bought ready weaned to his hand, and, indeed, at four months old, at which age, if he be in good condition, he will eat any-thing that an old hog will eat. He will graze, eat cabbage leaves, and almost the stumps. Swedish turnip tops or roots, and such things, with a little wash, will keep him along in very good growing order. I have now to speak of the time of purchasing, the manner of keeping, of fatting, killing, and curing; but these I must reserve till my next Number.

NO. VI.

KEEPING PIGS--(CONTINUED.)

143. As in the case of cows so in that of pigs, much must depend upon the situation of the cottage; because all pigs will graze; and therefore, on the skirts of forests or commons, a couple or three pigs may be kept, if the family be considerable; and especially if the cottager brew his own beer, which will give him grains to assist the wash. Even in lanes, or on the sides of great roads, a pig will find a good part of his food from May to November; and if he be yoked, the occupiers of the neighbourhood must be churlish and brutish indeed, if they give the owner any annoyance.

144. Let me break off here for a moment to point out to my readers the truly excellent conduct of Lord WINCHILSEA and Lord STANHOPE, who, as I read, have taken great pains to make the labourers on their estates comfortable, by allotting to each a piece of ground sufficient for the keeping of a cow. I once, when I lived at Botley, proposed to the copyholders and other farmers in my neighbourhood, that we should petition the Bishop of Winchester, who was lord of the manors thereabouts, to grant titles to all the numerous persons called trespassers on the wastes; and also to give titles to others of the poor parishioners, who were willing to make, on the skirts of the wastes, enclosures not exceeding an acre each. This I am convinced, would have done a great deal towards relieving the parishes, then greatly burdened by men out of work. This would have been better than digging holes one day to fill them up the next. Not a single man would agree to my proposal! One, a bullfrog farmer (now, I hear, pretty well sweated down,) said it would only make them saucy! And one, a true disciple of Malthus, said, that to facilitate their rearing of children was a harm! This man had, at the time, in his own occupation, land that had formerly been six farms, and he had, too, ten or a dozen children. I will not mention names; but this farmer will now, perhaps, have occasion to call to mind what I told him on that day, when his opposition, and particularly the ground of it, gave me the more pain, as he was a very industrious, civil, and honest man. Never was there a greater mistake than to suppose that men are made saucy and idle by just and kind treatment. Slaves are always lazy and saucy; nothing but the lash will extort from them either labour or respectful deportment. I never met with a saucy Yankee (New Englander) in my life. Never servile; always civil. This must necessarily be the character of freemen living in a state of competence. They have nobody to envy; nobody to complain of; they are in good humour with mankind. It must, however, be confessed, that very little, comparatively speaking, is to be accomplished by the individual efforts even of benevolent men like the two noblemen before mentioned. They have a strife to maintain against the general tendency of the national state of things. It is by general and indirect means, and not by partial and direct and positive regulations, that so great a good as that which they generously aim at can be accomplished. When we are to see such means adopted, God only knows; but, if much longer

delayed, I am of opinion, that they will come too late to prevent something very much resembling a dissolution of society.

145. The cottager's pig should be bought in the spring, or late in winter; and being then four months old, he will be a year old before killing time; for it should always be borne in mind, that this age is required in order to insure the greatest quantity of meat from a given quantity of food. If a hog be more than a year old, he is the better for it. The flesh is more solid and more nutritious than that of a young hog, much in the same degree that the mutton of a full-mouthed wether is better than that of a younger wether. The pork or bacon of young hogs, even if fatted on corn, is very apt to boil out, as they call it; that is to say, come out of the pot smaller in bulk than it goes in. When you begin to fat, do it by degrees, especially in the case of hogs under a year old. If you feed high all at once, the hog is apt to surfeit, and then a great loss of food takes place. Peas, or barley-meal is the food; the latter rather the best, and does the work quicker. Make him quite fat by all means. The last bushel, even if he sit as he eat, is the most profitable. If he can walk two hundred yards at a time, he is not well fatted. Lean bacon is the most wasteful thing that any family can use. In short, it is uneatable, except by drunkards, who want something to stimulate their sickly appetite. The man who cannot live on solid fat bacon, well-fed and well-cured, wants the sweet sauce of labour, or is fit for the hospital. But, then, it must be bacon, the effect of barley or peas, (not beans,) and not of whey, potatoes, or messes of any kind. It is frequently said, and I know that even farmers say it, that bacon, made from corn, costs more than it is worth! Why do they take care to have it then? They know better. They know well, that it is the very cheapest they can have; and they, who look at both ends and both sides of every cost, would as soon think of shooting their hogs as of fatting them on messes; that is to say, for their own use, however willing they might now-and-then be to regale the Londoners with a bit of potato-pork.

146. About Christmas, if the weather be coldish, is a good time to kill. If the weather be very mild, you may wait a little longer; for the hog cannot be too fat. The day before killing he should have no food. To kill a hog nicely is so much of a profession, that it is better to pay a shilling for having it done, than to stab and hack and tear the carcass about. I shall not speak of pork; for I would by no means recommend it. There are two ways of going to work to make bacon; in the one you take off the hair by scalding. This is the practice in most parts of England, and all over America. But the Hampshire way, and the best way, is to burn the hair off. There is a great deal of difference in the consequences. The first method slackens the skin, opens all the pores of it, makes it loose and flabby by drawing out the roots of the hair. The second tightens the skin in every part, contracts all the sinews and veins in the skin, makes the flitch a solider thing, and the skin a better protection to the meat. The taste of the meat is very different from that of a scalded hog; and to this chiefly it was that Hampshire bacon owed its reputation for excellence. As the hair is to be burnt off it must be dry, and care must be taken, that the hog be kept on dry litter of some sort the day previous to killing. When killed he is laid upon a narrow bed of straw, not wider than his carcass, and only two or three inches thick. He is then covered all over thinly with straw, to which, according as the wind may be, the fire is put at one end. As the straw burns, it burns the hair. It requires two or three coverings and burnings, and care is taken, that the skin be not in any part burnt, or parched. When the hair is all burnt off close, the hog is scraped clean, but never touched with water. The upper side being finished, the hog is turned over, and the other side is treated in like manner. This work should always be done before

day-light; for in the day-light you cannot so nicely discover whether the hair be sufficiently burnt off. The light of the fire is weakened by that of the day. Besides, it makes the boys get up very early for once at any rate, and that is something; for boys always like a bonfire.

147. The inwards are next taken out, and if the wife be not a slattern, here, in the mere offal, in the mere garbage, there is food, and delicate food too, for a large family for a week; and hog's puddings for the children, and some

for neighbours' children, who come to play with them; for these things are by no means to be overlooked, seeing that they tend to the keeping alive of that affection in children for their parents, which, later in life, will be found absolutely necessary to give effect to wholesome precept, especially when opposed to the boisterous passions of youth.

148. The butcher, the next day, cuts the hog up; and then the house is filled with meat! Souse, griskins, blade-bones, thigh-bones, spare-ribs, chines, belly-pieces, cheeks, all coming into use one after the other, and the last of the latter not before the end of about four or five weeks. But about this time, it is more than possible that the Methodist parson will pay you a visit. It is remarked in America, that these gentry are attracted by the squeaking of the pigs, as the fox is by the cackling of the hen. This may be called slander; but I will tell you what I did know to happen. A good honest careful fellow had a spare-rib, on which he intended to sup with his family after a long and hard day's work at coppice-cutting. Home he came at dark with his two little boys, each with a nitch of wood that they had carried four miles, cheered with the thought of the repast that awaited them. In he went, found his wife, the Methodist parson, and a whole troop of the sisterhood, engaged in prayer, and on the table lay scattered the clean-polished bones of the spare-rib! Can any reasonable creature believe, that, to save the soul, God requires us to give up the food necessary to sustain the body? Did Saint Paul preach this? He, who, while he spread the gospel abroad, worked himself, in order to have it to give to those who were unable to work? Upon what, then, do these modern saints; these evangelical gentlemen, found their claim to live on the labour of others.

149. All the other parts taken away, the two sides that remain, and that are called flitches, are to be cured for bacon. They are first rubbed with salt on their insides, or flesh sides, then placed, one on the other, the flesh sides uppermost, in a salting trough which has a gutter round its edges to drain away the brine; for, to have sweet and fine bacon, the flitches must not lie sopping in brine; which gives it that sort of taste which barrel-pork and sea-jonk have, and than which nothing is more villanous. Every one knows how different is the taste of fresh, dry salt, from that of salt in a dissolved state. The one is savoury, the other nauseous. Therefore, change the salt often. Once in four or five days. Let it melt, and sink in; but let it not lie too long. Change the flitches. Put that at bottom which was first put on the top. Do this a couple of times. This mode will cost you a great deal more in salt, or rather in taxes, than the sopping mode; but without it, your bacon will not be sweet and fine, and will not keep so well. As to the time required for making the flitches sufficiently salt, it depends on circumstances; the thickness of the flitch, the state of the weather, the place wherein the salting is going on. It takes a longer time for a thick than for a thin flitch; it takes longer in dry, than in damp weather; it takes longer in a dry than in a damp place. But for the flitches of a hog of twelve score, in weather not very dry or very damp, about six weeks may do; and as yours is to be fat, which receives little injury from over-salting, give time enough; for you are to have bacon till Christmas comes again. The place

for salting should, like a dairy, always be cool, but always admit of a free circulation of air: confined air, though cool, will taint meat sooner than the mid-day sun accompanied with a breeze. Ice will not melt in the hottest sun so soon as in a close and damp cellar. Put a lump of ice in cold water, and one of the same size before a hot fire, and the former will dissolve in half the time that the latter will. Let me take this occasion of observing, that an ice-house should never be under ground, or under the shade of trees. That the bed of it ought to be three feet above the level of the ground; that this bed ought to consist of something that will admit the drippings to go instantly off; and that the house should stand in a place open to the sun and air. This is the way they have the ice-houses under the burning sun of Virginia; and here they keep their fish and meat as fresh and sweet as in winter, when at the same time neither will keep for twelve hours, though let down to the depth of a hundred feet in a well. A Virginian, with some poles and straw, will stick up an ice-house for ten dollars, worth a dozen of those ice-houses, each of which costs our men of taste as many scores of pounds. It is very hard to imagine, indeed, what any one should want ice for, in a country like this, except for clodpole boys to slide upon, and to drown cockneys in skaiting-time; but if people must have ice in summer, they may as well go a right way as a wrong way to get it.

150. However, the patient that I have at this time under my hands wants nothing to cool his blood, but something to warm it, and, therefore, I will get back to the flitches of bacon, which are now to be smoked; for smoking is a great deal better than merely drying, as is the fashion in the dairy countries in the West of England. When there were plenty of farm-houses there were plenty of places to smoke bacon in; since farmers have lived in gentleman's houses, and the main part of the farm-houses have been knocked down, these places are not so plenty. However, there is scarcely any neighbourhood without a chimney left to hang bacon up in. Two precautions are necessary: first, to hang the flitches where no rain comes down upon them: second, not to let them be so near the fire as to melt.

These precautions taken, the next is, that the smoke must proceed from wood, not turf, peat, or coal. Stubble or litter might do; but the trouble would be great. Fir, or deal, smoke is not fit for the purpose. I take it, that the absence of wood, as fuel, in the dairy countries, and in the North, has led to the making of pork and dried bacon. As to the time that it requires to smoke a flitch, it must depend a good deal upon whether there be a constant fire beneath, and whether the fire be large or small. A month may do, if the fire be pretty constant, and such as a farm-house fire usually is. But over smoking, or, rather, too long hanging in the air, makes the bacon rust. Great attention should, therefore, be paid to this matter. The flitch ought not be dried up to the hardness of a board, and yet it ought to be perfectly dry.

Before you hang it up, lay it on the floor, scatter the flesh-side pretty thickly over with bran, or with some fine saw-dust other than that of deal or fir. Rub it on the flesh, or pat it well down upon it. This keeps the smoke from getting into the little openings, and makes a sort of crust to be dried on; and, in short, keeps the flesh cleaner than it would otherwise be.

151. To keep the bacon sweet and good, and free from nasty things that they call hoppers; that is to say, a sort of skipping maggots, engendered by a fly which has a great relish for bacon: to provide against this mischief, and also to keep the bacon from becoming rusty, the Americans, whose country is so hot in summer, have two methods. They smoke no part of the hog except the hams, or gammons. They cover these with coarse linen cloth such as the finest

hop-bags are made of, which they sew neatly on. They then white-wash the cloth all over with lime white-wash, such as we put on walls, their lime being excellent stone-lime. They give the ham four or five washings, the one succeeding as the former gets dry; and in the sun, all these washings are put on in a few hours. The flies cannot get through this; and thus the meat is preserved from them. The other mode, and that is the mode for you, is, to sift fine some clean and dry wood-ashes. Put some at the bottom of a box, or chest, which is long enough to hold a flitch of bacon. Lay in one flitch; then put in more ashes; then the other flitch; and then cover this with six or eight inches of the ashes. This will effectually keep away all flies; and will keep the bacon as fresh and good as when it came out of the chimney, which it will not be for any great length of time, if put on a rack, or kept hung up in the open air. Dust, or even sand, very, very dry, would, perhaps, do as well. The object is not only to keep out the flies, but the air. The place where the chest, or box, is kept, ought to be dry; and, if the ashes should get damp (as they are apt to do from the salts they contain,) they should be put in the fire-place to dry, and then be put back again. Peat-ashes, or turf-ashes, might do very well for this purpose. With these precautions, the bacon will be as good at the end of the year as on the first day; and it will keep two, and even three years, perfectly good, for which, however, there can be no necessity.

152. Now, then, this hog is altogether a capital thing. The other parts will be meat for about four or five weeks. The lard, nicely put down, will last a long while for all the purposes for which it is wanted. To make it keep well there should be some salt put into it. Country children are badly brought up if they do not like sweet lard spread upon bread, as we spread butter. Many a score hunches of this sort have I eaten, and I never knew what poverty was. I have eaten it for luncheon at the houses of good substantial farmers in France and Flanders. I am not now frequently so hungry as I ought to be; but I should think it no hardship to eat sweet lard instead of butter. But, now-a-days, the labourers, and especially the female part of them, have fallen into the taste of niceness in food and finery in dress; a quarter of a bellyful and rags are the consequence. The food of their choice is high-priced, so that, for the greater part of their time, they are half-starved. The dress of their choice is showy and flimsy, so that, to-day, they are ladies, and to-morrow ragged as sheep with the scab. But has not Nature made the country girls as pretty as ladies? Oh, yes! (bless their rosy cheeks and white teeth!) and a great deal prettier too! But are they less pretty, when their dress is plain and substantial, and when the natural presumption is, that they have smocks as well as gowns, than they are when drawn off in the frail fabric of Sir Robert Peel,[9] "where tawdry colours strive with dirty white," exciting violent suspicions that all is not as it ought to be nearer the skin, and calling up a train of ideas extremely hostile to that sort of feeling which every lass innocently and commendably wishes to awaken in her male beholders? Are they prettiest when they come through the wet and dirt safe and neat; or when their draggled dress is plastered to their backs by a shower of rain? However, the fault has not been theirs, nor that of their parents. It is the system of managing the affairs of the nation. This system has made all flashy and false, and has put all things out of their place. Pomposity, bombast, hyperbole, redundancy, and obscurity, both in speaking and in writing; mock-delicacy in manners; mock-liberality, mock-humanity, and mock-religion. Pitt's false money, Peel's flimsy dresses, Wilberforce's potatoe diet, Castlereagh's and Mackintosh's oratory, Walter Scott's poems, Walter's and Stoddart's[10] paragraphs, with all the bad taste and baseness and hypocrisy which they spread over this country; all have arisen, grown, branched out, bloomed, and borne together; and we are now beginning to taste of their fruit. But, as the fat of the adder is, as is said, the antidote to its sting; so in the Son of the great worker of

Spinning-Jennies, we have, thanks to the Proctors and Doctors of Oxford, the author of that Bill, before which this false, this flashy, this flimsy, this rotten system will dissolve as one of his father's pasted calicoes does at the sight of the washing-tub.

153. "What," says the cottager, "has all this to do with hogs and bacon?" Not directly with hogs and bacon, indeed; but it has a great deal to do, my good fellow with your affairs, as I shall, probably, hereafter more fully show, though I shall now leave you to the enjoyment of your flitches of bacon, which, as I before observed, will do ten thousand times more than any Methodist parson, or any other parson (except, of course, those of our church) to make you happy, not only in this world, but in the world to come. Meat in the house is a great source of harmony, a great preventer of the temptation to commit those things, which, from small beginnings, lead, finally, to the most fatal and atrocious results; and I hold that doctrine to be truly damnable, which teaches that God has made any selection, any condition relative to belief, which is to save from punishment those who violate the principles of natural justice.

154. Some other meat you may have; but, bacon is the great thing. It is always ready; as good cold as hot; goes to the field or the coppice conveniently; in harvest, and other busy times, demands the pot to be boiled only on a Sunday; has twice as much strength in it as any other thing of the same weight; and in short, has in it every quality that tends to make a labourer's family able to work and well off. One pound of bacon, such as that which I have described, is, in a labourer's family, worth four or five of ordinary mutton or beef, which are great part bone, and which, in short, are gone in a moment. But always observe, it is fat bacon that I am talking about. There will, in spite of all that can be done, be some lean in the gammons, though comparatively very little; and therefore you ought to begin at that end of the flitches; for, old lean bacon is not good.

155. Now, as to the cost. A pig (a spayed sow is best) bought in March four months old, can be had now for fifteen shillings. The cost till fatting time is next to nothing to a Cottager; and then the cost, at the present price of corn, would, for a hog of twelve score, not exceed three pounds; in the whole four pounds five; a pot of poison a week bought at the public-house comes to twenty-six shillings of the money; and more than three times the remainder is generally flung away upon the miserable tea, as I have clearly shown in the First Number, at Paragraph 24. I have, indeed, there shown, that if the tea were laid aside, the labourer might supply his family well with beer all the year round, and have a fat hog of even fifteen score for the cost of the tea, which does him and can do him no good at all.

156. The feet, the cheeks, and other bone, being considered, the bacon and lard, taken together, would not exceed sixpence a pound. Irish bacon is "cheaper." Yes, lower-priced. But, I will engage that a pound of mine, when it comes out of the pot (to say nothing of the taste,) shall weigh as much as a pound and a half of Irish, or any dairy or slop-fed bacon, when that comes out of the pot. No, no: the farmers joke when they say, that their bacon costs them more than they could buy bacon for. They know well what it is they are doing; and besides, they always forget, or, rather, remember not to say, that the fatting of a large hog yields them three or four load of dung, really worth more than ten or fifteen of common yard dung. In short, without hogs, farming could not go on; and it never has gone on in any country in the world. The hogs are the great stay of the whole concern. They are much in small space; they make no show, as

flocks and herds do; but with out them, the cultivation of the land would be a poor, a miserably barren concern.

SALTING MUTTON AND BEEF.

157. VERY FAT Mutton may be salted to great advantage, and also smoked, and may be kept thus a long while. Not the shoulders and legs, but the back of the sheep. I have never made any flitch of sheep-bacon; but I will; for there is nothing like having a store of meat in a house. The running to the butchers daily is a ridiculous thing. The very idea of being fed, of a family being fed, by daily supplies, has something in it perfectly tormenting. One half of the time of a mistress of a house, the affairs of which are carried on in this way, is taken up in talking about what is to be got for dinner, and in negotiations with the butcher. One single moment spent at table beyond what is absolutely necessary, is a moment very shamefully spent; but, to suffer a system of domestic economy, which unnecessarily wastes daily an hour or two of the mistress's time in hunting for the provision for the repast, is a shame indeed; and when we consider how much time is generally spent in this and in equally absurd ways, it is no wonder that we see so little performed by numerous individuals as they do perform during the course of their lives.

158. Very fat parts of Beef may be salted and smoked in a like manner. Not the lean; for that is a great waste, and is, in short, good for nothing. Poor fellows on board of ships are compelled to eat it, but it is a very bad thing.

NO. VII.

BEES, FOWLS, &C. &C.

159. I now proceed to treat of objects of less importance than the foregoing, but still such as may be worthy of great attention. If all of them cannot be

expected to come within the scope of a labourer's family, some of them must, and others may: and it is always of great consequence, that children be brought up to set a just value upon all useful things, and especially upon all living things; to know the utility of them: for, without this, they never, when grown up, are worthy of being entrusted with the care of them. One of the greatest, and, perhaps, the very commonest, fault of servants, is their inadequate care of animals committed to their charge. It is a well-known saying that "the master's eye makes the horse fat," and the remissness to which this alludes, is generally owing to the servant not having been brought up to feel an interest in the well-being of animals.

BEES.

160. It is not my intention to enter into a history of this insect about which so much has been written, especially by the French naturalists. It is the useful that I shall treat of, and that is done in not many words. The best hives are those made of clean unblighted rye-straw. Boards are too cold in England. A swarm should always be put into a new hive, and the sticks should be new that are put into the hive for the bees to work on; for, if the hive be old, it is not so wholesome, and a thousand to one but it contain the embryos of moths and other insects injurious to bees. Over the hive itself there should be a cap of thatch, made also of clean rye straw; and it should not only be new when first put on the hive; but a new one should be made to supply the place of the former one every three or four months; for when the straw begins to get rotten, as it soon does, insects breed in it, its smell is bad, and its effect on the bees is dangerous.

161. The hive should be placed on a bench, the legs of which mice and rats cannot creep up. Tin round the legs is best. But even this will not keep down ants, which are mortal enemies of bees. To keep these away, if you find them infest the hive, take a green stick and twist it round in the shape of a ring to lay on the ground round the leg of the bench, and at a few inches from it; and cover this stick with tar. This will keep away the ants. If the ants come from one home, you may easily trace them to it; and when you have found it, pour boiling water on it in the night, when all the family are at home.

This is the only effectual way of destroying ants, which are frequently so troublesome. It would be cruel to cause this destruction, if it were not necessary to do it, in order to preserve the honey, and indeed the bees too.

162. Besides the hive and its cap, there should be a sort of shed, with top, back, and ends, to give additional protection in winter; though in summer hives may be kept too hot, and in that case the bees become sickly and the produce becomes light. The situation of the hive is to face the South-east; or, at any rate, to be sheltered from the North and the West. From the North always, and from the West in winter. If it be a very dry season in summer, it contributes greatly to the success of the bees, to place clear water near their home, in a thing that they can conveniently drink out of; for if they have to go a great way for drink, they have not much time for work.

163. It is supposed that bees live only a year; at any rate it is best never to keep the same stall, or family, over two years, except you want to increase your number of hives. The swarm of this summer should always be taken in the autumn of next year. It is whimsical to save the bees when you take the honey. You must feed them; and, if saved, they will die of old age before the next fall; and though young ones will supply the place of the dead, this is nothing like a good swarm put up during the summer.

164. As to the things that bees make their collections from, we do not, perhaps, know a thousandth part of them; but of all the blossoms that they seek eagerly that of the Buck-wheat stands foremost. Go round a piece of this grain just towards sunset, when the buck-wheat is in bloom, and you will see the air filled with bees going home from it in all directions. The buck-wheat, too, continues in bloom a long while; for the grain is dead ripe on one part of the plant, while there are fresh blossoms coming out on the other part.

165. A good stall of bees, that is to say, the produce of one, is always worth about two bushels of good wheat. The cost is nothing to the labourer. He must be a stupid countryman indeed who cannot make a bee-hive; and a lazy one indeed if he will not, if he can. In short, there is nothing but care demanded; and there are very few situations in the country, especially in the south of England, where a labouring man may not have half a dozen stalls of bees to take every year. The main things are to keep away insects, mice, and birds, and especially a little bird called the bee-bird; and to keep all clean and fresh as to the hives and coverings. Never put a swarm into an old hive. If wasps, or hornets, annoy you, watch them home in the day time; and in the night kill them by fire, or by boiling water. Fowls should not go where bees are, for they eat them.

166. Suppose a man get three stalls of bees in a year. Six bushels of wheat give him bread for an eighth part of the year. Scarcely any thing is a greater misfortune than shiftlessness. It is an evil little short of the loss of eyes or of limbs.

GEESE.

167. They can be kept to advantage only where there are green commons, and there they are easily kept; live to a very great age; and are amongst the hardiest animals in the world. If well kept, a goose will lay a hundred eggs in a year. The French put their eggs under large hens of common fowls, to each of which they give four or five eggs; or under turkies, to which they give nine or ten goose-eggs. If the goose herself sit, she must be well and regularly fed, at, or near to, her nest. When the young ones are hatched, they should be kept in a warm place for about four days, and fed on barley-meal, mixed, if possible, with milk; and then they will begin to graze.

Water for them, or for the old ones to swim in, is by no means necessary, nor, perhaps, ever even useful. Or, how is it, that you see such fine flocks of fine geese all over Long Island (in America) where there is scarcely such a thing as a pond or a run of water?

168. Geese are raised by grazing; but to fat them something more is required. Corn of some sort, or boiled Swedish turnips. Some corn and some raw Swedish turnips, or carrots, or white cabbages, or lettuces, make the best fatting. The modes that are resorted to by the French for fatting geese, nailing them down by their webs, and other acts of cruelty, are, I hope, such as Englishmen will never think of. They will get fat enough without the use of any of these unfeeling means being employed. He who can deliberately inflict torture upon an animal, in order to heighten the pleasure his palate is to receive in eating it, is an abuser of the authority which God has given him, and is, indeed, a tyrant in his heart. Who would think himself safe, if at the mercy of such a man? Since the first edition of this work was published, I have had a good deal of experience with regard to geese. It is a very great error to suppose that what is called a Michaelmas goose is the thing. Geese are, in general, eaten at the age when they are called green geese; or after they have got their full and entire growth, which is not until the latter part of October. Green geese are tasteless squabs; loose flabby things; no rich taste in them; and, in short, a very indifferent sort of dish. The full-grown goose has solidity in it; but it is hard, as well as solid; and in place of being rich, it is strong. Now, there is a middle course to take; and if you take this course, you produce the finest birds of which we can know any thing in England. For three years, including the present year, I have had the finest geese that I ever saw, or ever heard of. I have bought from twenty to thirty every one of these years. I buy them off the common late in June, or very early in July. They have cost me from two shillings to three shillings each, first purchase. I bring the flock home, and put them in a pen, about twenty feet square, where I keep them well littered with straw, so as for them not to get filthy. They have one trough in which I give them dry oats, and they have another trough where they have constantly plenty of clean water. Besides these, we give them, two or three times a day, a parcel of lettuces out of the garden. We give them such as are going to seed generally; but the better the lettuces are, the better the geese. If we have no lettuces to spare, we give them cabbages, either loaved or not loaved; though, observe, the white cabbage as well as the white lettuce, that is to say, the loaved cabbage and lettuce, are a great deal better than those that are not loaved. This is the food of my geese. They thrive exceedingly upon this food. After we have had the flock about ten days, we begin to kill, and we proceed once or twice a week till about the middle of October, sometimes later. A great number of persons who have eaten of these geese have all declared that they did not imagine that a goose could be brought to be so good a bird. These geese are altogether different from the hard, strong things that come out of the stubble fields, and equally different from the flabby things called a green goose. I should think that the cabbages or lettuces perform half the work of keeping and fatting my geese; and these are things that really cost nothing. I should think that the geese, upon an average, do not consume more than a shilling's worth of oats each. So that we have these beautiful geese for about four shillings each. No money will buy me such a goose in London; but the thing that I can get nearest to it, will cost me seven shillings. Every gentleman has a garden. That garden has, in the month of July, a wagon-load, at least, of lettuces and cabbages to throw away. Nothing is attended with so little trouble as these geese. There is hardly any body near London that has not room for the purposes here mentioned. The reader will be apt to exclaim, as my friends very often do, "Cobbett's Geese are all Swans." Well, better that way than not to be pleased with what one has. However, let gentlemen try this method

of fatting geese. It saves money, mind, at the same time. Let them try it; and if any one, who shall try it, shall find the effect not to be that which I say it is, let him reproach me publicly with being a deceiver. The thing is no invention of mine. While I could buy a goose off the common for half-a-crown, I did not like to give seven shillings for one in London, and yet I wished that geese should not be excluded from my house. Therefore I bought a flock of geese, and brought them home to Kensington. They could not be eaten all at once. It was necessary, therefore, to fix upon a mode of feeding them. The above mode was adopted by my servant, as far as I know, without any knowledge of mine; but the very agreeable result made me look into the matter; and my opinion, that the information will be useful to many persons, at any rate, is sufficient to induce me to communicate it to my readers.

DUCKS.

169. No water, to swim in, is necessary to the old, and is injurious to the very young. They never should be suffered to swim (if water be near) till more than a month old. The old duck will lay, in the year, if well kept, ten dozen of eggs; and that is her best employment; for common hens are the best mothers. It is not good to let young ducks out in the morning to eat slugs and worms; for, though they like them, these things kill them if they eat a great quantity. Grass, corn, white cabbages, and lettuces, and especially buck-wheat, cut, when half ripe, and flung down in the haulm. This makes fine ducks. Ducks will feed on garbage and all sorts of filthy things; but their flesh is strong, and bad in proportion. They are, in Long Island, fatted upon a coarse sort of crab, called a horse-foot fish, prodigious quantities of which are cast on the shores. The young ducks grow very fast upon this, and very fat; but wo unto him that has to smell them when they come from the spit; and, as for eating them, a man must have a stomach indeed to do that!

170. When young, they should be fed upon barley-meal, or curds, and kept in a warm place in the night-time, and not let out early in the morning. They should, if possible, be kept from water to swim in. It always does them harm; and, if intended to be sold to be killed young, they should never go near ponds, ditches, or streams. When you come to fat ducks, you must take care that they get at no filth whatever. They will eat garbage of all sorts; they will suck down the most nauseous particles of all those substances which go for manure. A dead rat three parts rotten is a feast to them. For these reasons I should never eat any ducks, unless there were some mode of keeping them from this horrible food. I treat them precisely as I do my geese. I buy a troop when they are young, and put them in a pen, and feed them upon oats, cabbages, lettuces, and water, and have the place kept very clean. My ducks are, in consequence of this, a great deal more fine and delicate than any others that I know any-thing of.

TURKEYS.

171. These are flying things, and so are common fowls. But it may happen that a few hints respecting them may be of use. To raise turkeys in this chilly climate, is a matter of much greater difficulty than in the climates that give great warmth. But the great enemy to young turkeys (for old ones are hardy enough) is the wet. This they will endure in no climate; and so true is this, that, in America, where there is always "a wet spell" in April, the farmers' wives take care never to have a brood come out until that spell is passed. In England, where the wet spells

come at haphazard, the first thing is to take care that young turkeys never go out, on any account, except in dry weather, till the dew be quite off the ground; and this should be adhered to till they get to be of the size of an old partridge, and have their backs well covered with feathers. And, in wet weather, they should be kept under cover all day long.

172. As to the feeding of them, when young, various nice things are recommended. Hard eggs chopped fine, with crumbs of bread, and a great many other things; but that which I have seen used, and always with success, and for all sorts of young poultry, is milk turned into curds. This is the food for young poultry of all sorts. Some should be made fresh every day; and if this be done, and the young turkeys kept warm, and especially from wet, not one out of a score will die. When they get to be strong, they may have meal and grain, but still they always love the curds.

173. When they get their head feathers they are hardy enough; and what they then want is room to prowl about. It is best to breed them under a common hen; because she does not ramble like a hen-turkey; and it is a very curious thing that the turkeys bred up by a hen of the common fowl, do not themselves ramble much when they get old; and for this reason, when they buy turkeys for stock, in America, (where there are such large woods, and where the distant rambling of turkeys is inconvenient,) they always buy such as have been bred under the hens of the common fowl; than which a more complete proof of the great powers of habit is, perhaps, not to be found. And ought not this to be a lesson to fathers and mothers of families? Ought not they to consider that the habits which they give their children are to stick by those children during their whole lives?

174. The hen should be fed exceedingly well, too, while she is sitting and after she has hatched; for though she does not give milk, she gives heat; and, let it be observed, that as no man ever yet saw healthy pigs with a poor sow, so no man ever saw healthy chickens with a poor hen. This is a matter much too little thought of in the rearing of poultry; but it is a matter of the greatest consequence. Never let a poor hen sit; feed the hen well while she is sitting, and feed her most abundantly when she has young ones; for then her labour is very great; she is making exertions of some sort or other during the whole twenty-four hours; she has no rest; is constantly doing something or other to provide food or safety for her young ones.

175. As to fatting turkeys, the best way is, never to let them be poor. Cramming is a nasty thing, and quite unnecessary. Barley-meal, mixed with skim-milk, given to them, fresh and fresh, will make them fat in a short time, either in a coop, in a house, or running about. Boiled carrots and Swedish turnips will help, and it is a change of sweet food. In France they sometimes pick turkeys alive, to make them tender; of which I shall only say, that the man that can do this, or order it to be done, ought to be skinned alive himself.

FOWLS.

176. These are kept for two objects; their flesh and their eggs. As to rearing them, every thing said about rearing turkeys is applicable here. They are best fatted, too, in the same manner. But, as to laying-hens, there are some means to be used to secure the use of them in winter. They ought not to be old hens. Pullets, that is, birds hatched in the foregoing spring, are, perhaps, the best. At any rate, let them not be more than two years old. They should be kept in a warm place,

and not let out, even in the day-time, in wet weather; for one good sound wetting will keep them back for a fortnight. The dry cold, even in the severest cold, if dry, is less injurious than even a little wet in winter-time. If the feathers get wet, in our climate, in winter, or in short days, they do not get dry for a long time; and this it is that spoils and kills many of our fowls.

177. The French, who are great egg-eaters, take singular pains as to the food of laying-hens in winter. They let them out very little, even in their fine climate, and give them very stimulating food; barley boiled, and given them warm; curds, buck-wheat, (which, I believe, is the best thing of all except curds;) parsley and other herbs chopped fine; leeks chopped in the same way; also apples and pears chopped very fine; oats and wheat cribbled; and sometimes they give them hemp-seed, and the seed of nettles; or dried nettles, harvested in summer, and boiled in the winter. Some give them ordinary food, and, once a day, toasted bread sopped in wine. White cabbages chopped up are very good in winter for all sorts of poultry.

178. This is taking a great deal of pains; but the produce is also great and very valuable in winter; for, as to preserved eggs, they are things to run from and not after. All this supposes, however, a proper hen-house, about which we, in England, take very little pains. The vermin, that is to say, the lice, that poultry breed, are the greatest annoyance. And as our wet climate furnishes them, for a great part of the year, with no dust by which to get rid of these vermin, we should be very careful about cleanliness in the hen-houses. Many a hen, when sitting, is compelled to quit her nest to get rid of the lice. They torment the young chickens. And, in short, are a great injury. The fowl-house should, therefore, be very often cleaned out; and sand, or fresh earth, should be thrown on the floor. The nest should not be on shelves, or on any-thing fixed; but little flat baskets, something like those that the gardeners have in the markets in London, and which they call sieves, should be placed against the sides of the house upon pieces of wood nailed up for the purpose. By this means the nests are kept perfectly clean, because the baskets are, when necessary, taken down, the hay thrown out, and the baskets washed; which cannot be done, if the nest be made in any-thing forming a part of the building. Besides this, the roosts ought to be cleaned every week, and the hay changed in the nests of laying-hens. It is good to fumigate the house frequently by burning dry herbs, juniper wood, cedar wood, or with brimstone; for nothing stands so much in need of cleanliness as a fowl-house, in order to have fine fowls and plenty of eggs.

179. The ailments of fowls are numerous, but they would seldom be seen, if the proper care were taken. It is useless to talk of remedies in a case where you have complete power to prevent the evil. If well fed, and kept perfectly clean, fowls will seldom be sick; and, as to old age, they never ought to be kept more than a couple or three years; for they get to be good for little as layers, and no teeth can face them as food.

180. It is, perhaps, seldom that fowls can be kept conveniently about a cottage; but when they can, three, four, or half a dozen hens to lay in winter, when the wife is at home the greater part of the time, are worth attention. They would require but little room, might be bought in November and sold in April, and six of them, with proper care, might be made to clear every week the price of a gallon of flour. If the labour were great, I should not think of it; but it is none; and I am for neglecting nothing in the way of pains in order to ensure a hot dinner every day in winter, when the man comes home from work. As to the fatting of fowls, information can be of no use to those who live in a cottage all their lives; but it may be of some use to those who

are born in cottages, and go to have the care of poultry at richer persons' houses. Fowls should be put to fat about a fortnight before they are wanted to be killed. The best food is barley-meal wetted with milk, but not wetted too much. They should have clear water to drink, and it should be frequently changed. Crammed fowls are very nasty things: but "barn-door" fowls, as they are called, are sometimes a great deal more nasty. Barn-door would, indeed, do exceedingly well; but it unfortunately happens that the stable is generally pretty near to the barn. And now let any gentleman who talks about sweet barn-door fowls, have one caught in the yard, where the stable is also. Let him have it brought in, killed, and the craw taken out and cut open. Then let him take a ball of horse-dung from the stable-door; and let his nose tell him how very small is the difference between the smell of the horse-dung, and the smell of the craw of his fowl. In short, roast the fowl, and then pull aside the skin at the neck, put your nose to the place, and you will almost think that you are at the stable door. Hence the necessity of taking them away from the barn-door a fortnight, at least, before they are killed. We know very well that ducks that have been fed upon fish, either wild ducks, or tame ducks, will scent a whole room, and drive out of it all those who have not pretty good constitutions. It must be so. Solomon says that all flesh is grass; and those who know any-thing about beef, know the difference between the effect of the grass in Herefordshire and Lincolnshire, and the effect of turnips and oil cake. In America they always take the fowls from the farm-yard, and shut them up a fortnight or three weeks before they be killed. One thing, however, about fowls ought always to be borne in mind. They are never good for any-thing when they have attained their full growth, unless they be capons or poullards. If the poulets be old enough to have little eggs in them, they are not worth one farthing; and as to the cocks of the same age, they are fit for nothing but to make soup for soldiers on their march, and they ought to be taken for that purpose.

PIGEONS.

181. A few of these may be kept about any cottage, for they are kept even in towns by labourers and artizans. They cause but little trouble. They take care of their own young ones; and they do not scratch, or do any other mischief in gardens. They want feeding with tares, peas, or small beans; and buck-wheat is very good for them. To begin keeping them, they must not have flown at large before you get them. You must keep them for two or three days, shut into the place which is to be their home; and then they may be let out, and will never leave you, as long as they can get proper food, and are undisturbed by vermin, or unannoyed exceedingly by lice.

182. The common dove-house pigeons are the best to keep. They breed oftenest, and feed their young ones best. They begin to breed at about nine months old, and if well kept, they will give you eight or nine pair in the year. Any little place, a shelf in the cow shed; a board or two under the eaves of the house; or, in short, any place under cover, even on the ground floor, they will sit and hatch and breed up their young ones in.

183. It is not supposed that there could be much profit attached to them; but they are of this use; they are very pretty creatures; very interesting in their manners; they are an object to delight children, and to give them the early habit of fondness for animals and of setting a value on them, which, as I have often had to observe before, is a very great thing. A considerable part of all the property of a nation consists of animals. Of course a proportionate part of the cares and labours of a people appertain to the breeding and bringing to perfection those animals; and,

if you consult your experience, you will find that a labourer is, generally speaking, of value in proportion as he is worthy of being intrusted with the care of animals. The most careless fellow cannot hurt a hedge or ditch; but to trust him with the team, or the flock, is another matter. And, mind, for the man to be trust-worthy in this respect, the boy must have been in the habit of being kind and considerate towards animals; and nothing is so likely to give him that excellent habit as his seeing, from his very birth, animals taken great care of, and treated with great kindness by his parents, and now-and-then having a little thing to call his own.

RABBITS.

184. In this case, too, the chief use, perhaps, is to give children those habits of which I have been just speaking. Nevertheless, rabbits are really profitable. Three does and a buck will give you a rabbit to eat for every three days in the year, which is a much larger quantity of food than any man will get by spending half his time in the pursuit of wild animals, to say nothing of the toil, the tearing of clothes, and the danger of pursuing the latter.

185. Every-body knows how to knock up a rabbit hutch. The does should not be allowed to have more than seven litters in a year. Six young ones to a doe is all that ought to be kept; and then they will be fine. Abundant food is the main thing; and what is there that a rabbit will not eat? I know of nothing green that they will not eat; and if hard pushed, they will eat bark, and even wood. The best thing to feed the young ones on when taken from the mother, is the carrot, wild or garden. Parsnips, Swedish turnips, roots of dandelion; for too much green or watery stuff is not good for weaning rabbits. They should remain as long as possible with the mother. They should have oats once a-day; and, after a time, they may eat any-thing with safety. But if you give them too much green at first when they are weaned, they rot as sheep do. A variety of food is a great thing; and, surely, the fields and gardens and hedges furnish this variety! All sorts of grasses, strawberry-leaves, ivy, dandelions, the hog-weed or wild parsnip, in root, stem, and leaves. I have fed working horses, six or eight in number, upon this plant for weeks together. It is a tall bold plant that grows in prodigious quantities in the hedges and coppices in some parts of England. It is the perennial parsnip. It has flower and seed precisely like those of the parsnip; and hogs, cows, and horses, are equally fond of it. Many a half-starved pig have I seen within a few yards of cart-loads of this pig-meat! This arises from want of the early habit of attention to such matters. I, who used to get hog-weed for pigs and for rabbits when a little chap, have never forgotten that the wild parsnip is good food for pigs and rabbits.

186. When the doe has young ones, feed her most abundantly with all sorts of greens and herbage and with carrots and the other things mentioned before, besides giving her a few oats once a-day. That is the way to have fine healthy young ones, which, if they come from the mother in good case, will very seldom die. But do not think, that because she is a small animal, a little feeding is sufficient! Rabbits eat a great deal more than cows or sheep in proportion to their bulk.

187. Of all animals rabbits are those that boys are most fond of. They are extremely pretty, nimble in their movements, engaging in their attitudes, and always completely under immediate control. The produce has not long to be waited for. In short, they keep an interest constantly alive in a little chap's mind; and they really cost nothing; for as to the oats, where is

the boy that cannot, in harvest-time, pick up enough along the lanes to serve his rabbits for a year? The care is all; and the habit of taking care of things is, of itself, a most valuable possession.

188. To those gentlemen who keep rabbits for the use of their family (and a very useful and convenient article they are,) I would observe, that when they find their rabbits die, they may depend on it, that ninety-nine times out of the hundred starvation is the malady. And particularly short feeding of the doe, while, and before she has young ones; that is to say, short feeding of her at all times; for, if she be poor, the young ones will be good for nothing. She will live being poor, but she will not, and cannot breed up fine young ones.

GOATS AND EWES.

189. In some places where a cow cannot be kept, a goat may. A correspondent points out to me, that a Dorset ewe or two might be kept on a common near a cottage to give milk; and certainly this might be done very well; but I should prefer a goat, which is hardier and much more domestic. When I was in the army, in New Brunswick, where, be it observed, the snow lies on the ground seven months in the year, there were many goats that belonged to the regiment, and that went about with it on shipboard and every-where else. Some of them had gone through nearly the whole of the American War. We never fed them. In summer they picked about wherever they could find grass; and in winter they lived on cabbage-leaves, turnip-peelings, potatoe-peelings, and other things flung out of the soldiers' rooms and huts. One of these goats belonged to me, and, on an average throughout the year, she gave me more than three half-pints of milk a day. I used to have the kid killed when a few days old; and, for some time, the goat would give nearly or quite, two quarts of milk a day. She was seldom dry more than three weeks in the year.

190. There is one great inconvenience belonging to goats; that is, they bark all young trees that they come near; so that, if they get into a garden, they destroy every thing. But there are seldom trees on commons, except such as are too large to be injured by goats; and I can see no reason against keeping a goat where a cow cannot be kept. Nothing is so hardy; nothing is so little nice as to its food. Goats will pick peelings out of the kennel and eat them. They will eat mouldy bread or biscuit; fusty hay, and almost rotten straw; furze-bushes, heath-thistles; and, indeed, what will they not eat, when they will make a hearty meal on paper, brown or white, printed on or not printed on, and give milk all the while! They will lie in any dog-hole. They do very well clogged, or stumped out. And, then, they are very healthy things into the bargain, however closely they may be confined. When sea voyages are so stormy as to kill geese, ducks, fowls, and almost pigs, the goats are well and lively; and when a dog of no kind can keep the deck for a minute, a goat will skip about upon it as bold as brass.

191. Goats do not ramble from home. They come in regularly in the evening, and if called, they come like dogs. Now, though ewes, when taken great care of, will be very gentle, and though their milk may be rather more delicate than that of the goat, the ewes must be fed with nice and clean food, and they will not do much in the milk-giving way upon a common; and, as to feeding them, provision must be made pretty nearly as for a cow. They will not endure confinement like goats; and they are subject to numerous ailments that goats know nothing of. Then the ewes are done by the time they are about six years old; for they then lose their teeth; whereas a goat will continue to breed and to give milk in abundance for a great many years. The

sheep is frightened at everything, and especially at the least sound of a dog. A goat, on the contrary, will face a dog, and if he be not a big and courageous one, beat him off.

192. I have often wondered how it happened that none of our labourers kept goats; and I really should be glad to see the thing tried. They are pretty creatures, domestic as a dog, will stand and watch, as a dog does, for a crumb of bread, as you are eating; give you no trouble in the milking; and I cannot help being of opinion, that it might be of great use to introduce them amongst our labourers.

CANDLES AND RUSHES.

193. We are not permitted to make candles ourselves, and if we were, they ought seldom to be used in a labourer's family. I was bred and brought up mostly by rush-light, and I do not find that I see less clearly than other people. Candles certainly were not much used in English labourers' dwellings in the days when they had meat dinners and Sunday coats. Potatoes and taxed candles seem to have grown into fashion together; and, perhaps, for this reason: that when the pot ceased to afford grease for the rushes, the potatoe-gorger was compelled to go to the chandler's shop for light to swallow the potatoes by, else he might have devoured peeling and all!

194. My grandmother, who lived to be pretty nearly ninety, never, I believe, burnt a candle in her house in her life. I know that I never saw one there, and she, in a great measure, brought me up. She used to get the meadow-rushes, such as they tie the hop-shoots to the poles with. She cut them when they had attained their full substance, but were still green. The rush at this age, consists of a body of pith with a green skin on it. You cut off both ends of the rush, and leave the prime part, which, on an average, may be about a foot and a half long. Then you take off all the green skin, except for about a fifth part of the way round the pith. Thus it is a piece of pith all but a little strip of skin in one part all the way up, which, observe, is necessary to hold the pith together all the way along.

195. The rushes being thus prepared, the grease is melted, and put in a melted state into something that is as long as the rushes are. The rushes are put into the grease; soaked in it sufficiently; then taken out and laid in a bit of bark taken from a young tree, so as not to be too large. This bark is fixed up against the wall by a couple of straps put round it; and there it hangs for the purpose of holding the rushes.

196. The rushes are carried about in the hand; but to sit by, to work by, or to go to bed by, they are fixed in stands made for the purpose, some of which are high to stand on the ground, and some low, to stand on a table. These stands have an iron port something like a pair of pliers to hold the rush in, and the rush is shifted forward from time to time, as it burns down to the thing that holds it.

197. Now these rushes give a better light than a common small dip-candle; and they cost next to nothing, though the labourer may with them have as much light as he pleases, and though, without them he must sit the far greater part of the winter evenings in the dark, even if he expend fifteen shillings a year in candles. You may do any sort of work by this light; and, if reading be your taste, you may read the foul libels, the lies and abuse, which are circulated gratis about me

by the "Society for promoting Christian Knowledge," as well by rush-light, as you can by the light of taxed candles; and, at any rate, you would have one evil less; for to be deceived and to pay a tax for the deception are a little too much for even modern loyalty openly to demand.

MUSTARD.

198. Why buy this, when you can grow it in your garden? The stuff you buy is half drugs; and is injurious to health. A yard square of ground, sown with common Mustard, the crop of which you would grind for use, in a little mustard-mill, as you wanted it, would save you some money, and probably save your life. Your mustard would look brown instead of yellow; but the former colour is as good as the latter: and, as to the taste, the real mustard has certainly a much better than that of the drugs and flour which go under the name of mustard. Let any one try it, and I am sure he will never use the drugs again. The drugs, if you take them freely, leave a burning at the pit of your stomach, which the real mustard does not.

DRESS, HOUSEHOLD GOODS, AND FUEL.

199. In Paragraph 152, I said, I think, enough to caution you, the English labourer, against the taste, now too prevalent, for fine and flimsy dress. It was, for hundreds of years, amongst the characteristics of the English people, that their taste was, in all matters, for things solid, sound, and good; for the useful, and decent, the cleanly in dress, and not for the showy. Let us hope that this may be the taste again; and let us, my friends, fear no troubles, no perils, that may be necessary to produce a return of that taste, accompanied with full bellies and warm backs to the labouring classes.

200. In household goods, the warm, the strong, the durable, ought always to be kept in view. Oak tables, bedsteads and stools, chairs of oak or of yew tree, and never a bit of miserable deal board. Things of this sort ought to last several lifetimes. A labourer ought to inherit from his great grandfather something besides his toil. As to bedding, and other things of that sort, all ought to be good in their nature, of a durable quality, and plain in their colour and form. The plates, dishes, mugs, and things of that kind, should be of pewter, or even of wood. Any-thing is better than crockery-ware. Bottles to carry a-field should be of wood. Formerly, nobody but the gypsies and mumpers, that went a hop-picking in the season, carried glass or earthen bottles. As to glass of any sort, I do not know what business it has in any man's house, unless he be rich enough to live on his means. It pays a tax, in many cases, to the amount of two-thirds of its cost. In short, when a house is once furnished with sufficient goods, there ought to be no renewal of hardly any part of them wanted for half an age, except in case of destruction by fire. Good management in this way leaves the man's wages to provide an abundance of good food and good raiment; and these are the things that make happy families; these are the things that make a good, kind, sincere, and brave people; not little pamphlets about "loyalty" and "content." A good man will be contented fast enough, if he be fed and clad sufficiently; but if a man be not well fed and clad, he is a base wretch to be contented.

201. Fuel should be, if possible, provided in summer, or at least some of it. Turf and peat must be got in summer, and some wood may. In the woodland countries, the next winter ought to be thought of in June, when people hardly know what to do with the fuelwood; and something should, if possible, be saved in the bark-harvest to get a part of the fuel for the next

winter. Fire is a capital article. To have no fire, or a bad fire, to sit by, is a most dismal thing. In such a state man and wife must be something out of the common way to be in good humour with each other, to say nothing of colds and other ailments which are the natural consequence of such misery. If we suppose the great Creator to condescend to survey his works in detail, what object can be so pleasing to him as that of the labourer, after his return from the toils of a cold winter day, sitting with his wife and children round a cheerful fire, while the wind whistles in the chimney and the rain pelts the roof? But, of all God's creation, what is so miserable to behold or to think of as a wretched, half-starved family creeping to their nest of flocks or straw, there to lie shivering, till sent forth by the fear of absolutely expiring from want?

HOPS.

202. I treated of them before; but before I conclude this little Work, it is necessary to speak of them again. I made a mistake as to the tax on the Hops. The positive tax is 2d. a pound, and I (in former editions) stated it at 4d. However, in all such cases, there falls upon the consumer the expenses attending the paying of the tax. That is to say, the cost of interest of capital in the grower who pays the tax, and who must pay for it, whether his hops be cheap or dear. Then the trouble it gives him, and the rules he is compelled to obey in the drying and bagging, and which cause him great expense. So that the tax on hops of our own English growth, may now be reckoned to cost the consumer about 3-1/4d. a pound.

YEAST.

203. Yeast is a great thing in domestic management. I have once before published a receipt for making yeast-cakes, I will do it again here.

204. In Long Island they make yeast-cakes. A parcel of these cakes is made once a year. That is often enough. And, when you bake, you take one of these cakes (or more according to the bulk of the batch) and with them raise your bread. The very best bread I ever ate in my life was lightened with these cakes.

205. The materials for a good batch of cakes are as follows:--3 ounces of good fresh Hops; 3-1/2 pounds of Rye Flour; 7 pounds of Indian Corn Meal; and one Gallon of Water.-- Rub the hops, so as to separate them. Put them into the water, which is to be boiling at the time. Let them boil half an hour. Then strain the liquor through a fine sieve into an earthen vessel. While the liquor is hot, put in the Rye-Flour; stirring the liquor well, and quickly, as the Rye-Flour goes into it. The day after, when it is working, put in the Indian Meal, stirring it well as it goes in. Before the Indian Meal be all in, the mess will be very stiff; and it will, in fact, be dough, very much of the consistence of the dough that bread is made of.--Take this dough; knead it well, as you would for pie-crust. Roll it out with a rolling-pin, as you roll out pie-crust, to the thickness of about a third of an inch. When you have it (or a part of it at a time) rolled out, cut it up into cakes with a tumbler glass turned upside down, or with something else that will answer the same purpose. Take a clean board (a tin may be better) and put the cakes to dry in the sun. Turn them every day; let them receive no wet; and they will become as hard as ship biscuit. Put them into a bag, or box, and keep them in a place perfectly free from damp. When you bake, take two cakes, of the thickness above-mentioned, and about 3 inches in diameter; put them into hot water, over-night, having cracked them first.

Let the vessel containing them stand near the fire-place all night. They will dissolve by the morning, and then you use them in setting your sponge (as it is called) precisely as you would use the yeast of beer.

206. There are two things which may be considered by the reader as obstacles. FIRST, where are we to get the Indian Meal? Indian Meal is used merely because it is of a less adhesive nature than that of wheat. White pea-meal, or even barley-meal, would do just as well. But SECOND, to dry the cakes, to make them (and quickly too, mind) as hard as ship biscuit (which is much harder than the timber of Scotch firs or Canada firs;) and to do this in the sun (for it must not be fire,) where are we, in this climate, to get the sun? In 1816 we could not; for, that year, melons rotted in the glazed frames and never ripened. But, in every nine summers out of ten, we have in June, in July, or in August, a fortnight of hot sun, and that is enough. Nature has not given us a peach-climate; but we get peaches. The cakes, when put in the sun, may have a glass sash, or a hand-light, put over them. This would make their birth hotter than that of the hottest open-air situation in America. In short to a farmer's wife, or any good housewife, all the little difficulties to the attainment of such an object would appear as nothing. The will only is required; and, if there be not that, it is useless to think of the attempt.

SOWING SWEDISH TURNIP SEED.

207. It is necessary to be a little more full than I have been before as to the manner of sowing this seed; and I shall make my directions such as to be applied on a small or a large scale.--Those that want to transplant on a large scale will, of course, as to the other parts of the business, refer to my larger work.--It is to get plants for transplanting that I mean to sow the Swedish Turnip Seed. The time for sowing must depend a little upon the nature of the situation and soil. In the north of England, perhaps early in April may be best; but, in any of these southern counties, any time after the middle of April and before the 10th of May, is quite early enough. The ground which is to receive the seed should be made very fine, and manured with wood-ashes, or with good compost well mixed with the earth. Dung is not so good; for it breeds the fly more; or, at least, I think so. The seed should be sown in drills an inch deep, made as pointed out under the head of Sowing in my book on Gardening. When deposited in the drills evenly but not thickly, the ground should be raked across the drills, so as to fill them up; and then the whole of the ground should be trodden hard, with shoes not nailed, and not very thick in the sole. The ground should be laid out in four-feet beds for the reasons mentioned in the "Gardener." When the seeds come up, thin the plants to two inches apart as soon as you think them clear from the fly; for, if left thicker, they injure each other even in this infant state. Hoe frequently between the rows even before thinning the plants; and when they are thinned, hoe well and frequently between them; for this has a tendency to make them strong; and the hoeing before thinning helps to keep off the fly. A rod of ground, the rows being eight inches apart, and plants two inches apart in the row, will contain about two thousand two hundred plants. An acre in rows four feet apart and the plants a foot apart in the row, will take about ten thousand four hundred and sixty plants. So that to transplant an acre, you must sow about five rods of ground. The plants should be kept very clean; and, by the last week in June, or first in July, you put them out. I have put them out (in England) at all times between 7th of June and middle of August. The first is certainly earlier than I like; and the very finest I ever grew in England, and the finest I ever saw for a large piece, were transplanted on the 14th of July. But one year with another, the

last week in June is the best time. For size of plants, manner of transplanting, intercultivation, preparing the land, and the rest, see "Year's Residence in America."

NO. VIII.

ON THE CONVERTING OF ENGLISH GRASS, AND GRAIN PLANTS CUT GREEN, INTO STRAW, FOR THE PURPOSE OF MAKING PLAT FOR HATS AND BONNETS.

KENSINGTON, MAY 30, 1823.

208. The foregoing Numbers have treated, chiefly, of the management of the affairs of a labourer's family, and more particularly of the mode of disposing of the money earned by the labour of the family. The present Number will point out what I hope may become an advantageous kind of labour. All along I have proceeded upon the supposition, that the wife and children of the labourer be, as constantly as possible, employed in work of some sort or other. The cutting, the bleaching, the sorting, and the platting of straw, seem to be, of all employments, the best suited to the wives and children of country labourers; and the discovery which I have made, as to the means of obtaining the necessary materials, will enable them to enter at once upon that employment.

209. Before I proceed to give my directions relative to the performance of this sort of labour, I shall give a sort of history of the discovery to which I have just alluded.

210. The practice of making hats, bonnets, and other things, of straw, is perhaps of very ancient date; but not to waste time in fruitless inquiries, it is very well known that, for many years past, straw coverings for the head have been greatly in use in England, in America, and, indeed, in almost all the countries that we know much of. In this country the manufacture was, only a few years ago, very flourishing; but it has now greatly declined, and has left in poverty and misery those whom it once well fed and clothed.

211. The cause of this change has been, the importation of the straw hats and bonnets from Italy, greatly superior, in durability and beauty, to those made in England. The plat made in England was made of the straw of ripened grain. It was, in general, split; but the main circumstance was, that it was made of the straw of ripened grain; while the Italian plat was made of the straw of grain, or grass, cut green. Now, the straw of ripened grain or grass is brittle; or, rather, rotten. It dies while standing, and, in point of toughness, the difference between it and straw from plants cut green is much about the same as the difference between a stick that has

died on the tree, and one that has been cut from the tree. But besides the difference in point of toughness, strength, and durability, there was the difference in beauty. The colour of the Italian plat was better; the plat was brighter; and the Indian straws, being small whole straws, instead of small straws made by the splitting of large ones, here was a roundness in them, that gave light and shade to the plat, which could not be given by our flat bits of straw.

212. It seems odd, that nobody should have set to work to find out how the Italians came by this fine straw. The importation of these Italian articles was chiefly from the port of LEGHORN; and therefore the bonnets imported were called Leghorn Bonnets. The straw manufacturers in this country seem to have made no effort to resist this invasion from Leghorn. And, which is very curious, the Leghorn straw has now began to be imported, and to be platted in this country. So that we had hands to plat as well as the Italians. All that we wanted was the same kind of straw that the Italians had: and it is truly wonderful that these importations from Leghorn should have gone on increasing year after year, and our domestic manufacture dwindling away at a like pace, without there having been any inquiry relative to the way in which the Italians got their straw! Strange, that we should have imported even straw from Italy, without inquiring whether similar straw could not be got in England! There really seems to have been an opinion, that England could no more produce this straw than it could produce the sugar-cane.

213. Things were in this state, when in 1821, a Miss WOODHOUSE, a farmer's daughter in CONNECTICUT, sent a straw-bonnet of her own making to the Society of Arts in London. This bonnet, superior in fineness and beauty to anything of the kind that had come from Leghorn, the maker stated to consist of a sort of grass of which she sent along with the bonnet some of the seeds. The question was, then, would these precious seeds grow and produce plants in perfection in England? A large quantity of the seed had not been sent: and it was therefore, by a member of the Society, thought desirable to get, with as little delay as possible, a considerable quantity of the seed.

214. It was in this stage of the affair that my attention was called to it. The member just alluded to applied to me to get the seed from America. I was of opinion that there could be no sort of grass in Connecticut that would not, and that did not, grow and flourish in England. My son JAMES, who was then at New-York, had instructions from me, in June 1821, to go to Miss WOODHOUSE, and to send me home an account of the matter. In September, the same year, I heard from him, who sent me an account of the cutting and bleaching, and also a specimen of the plat and grass of Connecticut. Miss WOODHOUSE had told the Society of Arts, that the grass used was the Poa Pratensis. This is the smooth-stalked meadow-grass. So that it was quite useless to send for seed. It was clear, that we had grass enough in England, if we could but make it into straw as handsome as that of Italy.

215. Upon my publishing an account of what had taken place with regard to the American Bonnet, an importer of Italian straw applied to me to know whether I would undertake to import American straw. He was in the habit of importing Italian straw, and of having it platted in this country; but having seen the bonnet of Miss WOODHOUSE, he was anxious to get the American straw. This gentleman showed me some Italian straw which he had imported, and as the seed heads were on, I could not see what plant it was. The gentleman who showed the straw to me, told me (and, doubtless, he believed) that the plant was one that would not grow in

England. I however, who looked at the straw with the eyes of a farmer, perceived that it consisted of dry oat, wheat, and rye plants, and of Bennet and other common grass plants.

216. This quite settled the point of growth in England. It was now certain that we had the plants in abundance; and the only question that remained to be determined was, Had we SUN to give to those plants the beautiful colour which the American and Italian straw had? If that colour were to be obtained by art, by any chemical applications, we could obtain it as easily as the Americans or the Italians; but, if it were the gift of the SUN solely, here might be a difficulty impossible for us to overcome. My experiments have proved that the fear of such difficulty was wholly groundless.

217. It was late in September 1821 that I obtained this knowledge, as to the kind of plants that produced the foreign straw. I could, at that time of the year, do nothing in the way of removing my doubts as to the powers of our Sun in the bleaching of grass; but I resolved to do this when the proper season for bleaching should return. Accordingly, when the next month of June came, I went into the country for the purpose. I made my experiments, and, in short, I proved to demonstration, that we had not only the plants, but the sun also, necessary for the making of straw, yielding in no respect to that of America or of Italy. I think that, upon the whole, we have greatly the advantage of those countries; for grass is more abundant in this country than in any other. It flourishes here more than in any other country. It is here in a greater variety of sorts; and for fineness in point of size, there is no part of the world which can equal what might be obtained from some of our downs, merely by keeping the land ungrazed till the month of July.

218. When I had obtained the straw, I got some of it made into plat. One piece of this plat was equal in point of colour, and superior in point of fineness, even to the plat of the bonnet, of Miss WOODHOUSE. It seemed, therefore, now to be necessary to do nothing more than to make all this well known to the country. As the SOCIETY OF ARTS had interested itself in the matter, and as I heard that, through its laudable zeal, several sowings of the foreign grass-seed had been made in England, I communicated an account of my experiments to that Society. The first communication was made by me on the 19th of February last, when I sent to the Society, specimens of my straw and also of the plat. Some time after this I attended a committee of the Society on the subject, and gave them a verbal account of the way in which I had gone to work.

219. The committee had, before this, given some of my straw to certain manufacturers of plat, in order to see what it would produce. These manufacturers, with the exception of one, brought such specimens of plat as to induce, at first sight, any one to believe that it was nonsense to think of bringing the thing to any degree of perfection! But, was it possible to believe this? Was it possible to believe that it could answer to import straw from Italy, to pay a twenty per cent. duty on that straw, and to have it platted here; and that it would not answer to turn into plat straw of just the same sort grown in England? It was impossible to believe this; but possible enough to believe, that persons now making profit by Italian straw, or plat, or bonnets, would rather that English straw should come to shut out the Italian and to put an end to the Leghorn trade.

220. In order to show the character of the reports of those manufacturers, I sent some parcels of straw into Hertfordshire, and got back, in the course of five days, fifteen specimens

of plat. These I sent to the Society of Arts on the 3d of April; and I here insert a copy of the letter which accompanied them.

TO THE SECRETARY OF THE SOCIETY OF ARTS. KENSINGTON, April 3, 1823.

SIR,--With this letter I send you sixteen specimens of plat, and also eight parcels of straw, in order to show the sorts that the plat is made out of. The numbers of the plat correspond with those of the straw; but each parcel of straw has two numbers attached to it, except in the case of the first number, which is the wheat straw. Of each kind of straw a parcel of the stoutest and a parcel of the smallest were sent to be platted; so that each parcel of the straw now sent, except that of the wheat, refers to two of the pieces of plat.

For instance, 2 and 3 of the plat is of the sort of straw marked 2 and 3; 4 and 12 of the plat is of the sort of straw marked 4 and 12; and so on. These parcels of straw are sent in order that you may know the kind of straw, or rather, of grass, from which the several pieces of plat have been made. This is very material; because it is by those parcels of straw that the kinds of grass are to be known.

The piece of plat No. 16 is American; all the rest are from my straw. You will see, that 15 is the finest plat of all. No. 7 is from the stout straws of the same kind as No. 15. By looking at the parcel of straw Nos. 7 and 15, you will see what sort of grass this is. The next, in point of beauty and fineness combined, are the pieces Nos. 13 and 8; and by looking at the parcel of straw, Nos. 13 and 8, you will see what sort of grass that is. Next comes 10 and 5, which are very beautiful too; and the sort of grass, you will see, is the common Bennet. The wheat, you see, is too coarse; and the rest of the sorts are either too hard or too brittle. I beg you to look at Nos. 10 and 5.

Those appear to me to be the thing to supplant the Leghorn. The colour is good, the straws work well, they afford a great variety of sizes, and they come from the common Bennet grass, which grows all over the kingdom, which is cultivated in all our fields, which is in bloom in the fair month of June, which may be grown as fine or as coarse as we please, and ten acres of which would, I dare say, make ten thousand bonnets. However, 7 and 15, and 8 and 13, are very good; and they are to be got in every part of the kingdom.

As to platters, it is to be too childish to believe that they are not to be got, when I could send off these straws, and get back the plat, in the course of five days. Far better work than this would have been obtained if I could have gone on the errand myself. What then will people not do, who regularly undertake the business for their livelihood?

I will, as soon as possible, send you an account of the manner in which I went to work with the grass. The card or plat, which I sent you some time ago, you will be so good as to give me back again some time; because I have now not a bit of the American plat left.

I am, Sir, your most humble and most obedient servant, WM. COBBETT.

221. I should observe, that these written communications, of mine to the Society, belong, in fact, to it, and will be published in its PROCEEDINGS, a volume of which comes

out every year; but, in this case, there would have been a year lost to those who may act in consequence of these communications being made public. The grass is to be got, in great quantities and of the best sorts, only in June and July; and the Society's volume does not come out till December. The Society has, therefore, given its consent to the making of the communications public through the means of this little work of mine.

222. Having shown what sort of plat could be produced from English grass-straw, I next communicated to the Society an account of the method which I pursued in the cutting and bleaching of the grass. The letter in

which I did this I shall here insert a copy of, before I proceed further. In the original the paragraphs were numbered from one to seventeen: they are here marked by letters, in order to avoid confusion, the paragraphs of the work itself being marked by numbers.

TO THE SECRETARY OF THE SOCIETY OF ARTS. KENSINGTON, April 14, 1823.

A.--SIR,--Agreeably to your request, I now communicate to you a statement of those particulars which you wished to possess, relative to the specimens of straw and of plat which I have at different times sent to you for the inspection of the Society.

B.--That my statement may not come too abruptly upon those members of the Society who have not had an opportunity of witnessing the progress of this interesting inquiry, I will take a short review of the circumstances which led to the making of my experiments.

C.--In the month of June, 1821, a gentleman, a member of the Society, informed me, by letter, that a Miss WOODHOUSE, a farmer's daughter, of Weathersfield, in Connecticut, had transmitted to the Society a

straw-bonnet of very fine materials and manufacture; that this bonnet (according to her account) was made from the straw of a sort of grass called poa pratensis; that it seemed to be unknown whether the same grass would grow in England; that it was desirable to ascertain whether this grass would grow in England; that, at all events, it was desirable to get from America some of the seed of this grass; and that, for this purpose, my informant, knowing that I had a son in America; addressed himself to me, it being his opinion that, if materials similar to those used by Miss WOODHOUSE could by any means be grown in England, the benefit to the nation must be considerable.

D.--In consequence of this application, I wrote to my son James, (then at New York,) directing him to do what he was able in order to cause success to the undertaking. On the receipt of my letter, in July, he went from New York to Weathersfield, (about a hundred and twenty miles;) saw Miss WOODHOUSE; made the necessary inquiries; obtained a specimen of the grass, and also of the plat, which other persons at Weathersfield, as well as Miss WOODHOUSE, were in the habit of making; and having acquired the necessary information as to cutting the grass and bleaching the straw, he transmitted to me an account of the matter; which account, together with his specimens of grass and plat, I received in the month of September.

E.--I was now, when I came to see the specimen of grass, convinced that Miss WOODHOUSE'S materials could be grown in England; a conviction which, if it had not been complete at once, would have been made complete immediately afterwards by the sight of a bunch of bonnet-straw imported from Leghorn, which straw was shown to me by the importer, and which I found to be that of two or three sorts of our common grass, and of oats, wheat, and rye.

F.--That the grass, or plants, could be grown in England was, therefore, now certain, and indeed that they were, in point of commonness, next to the earth itself. But before the grass could, with propriety, be called materials for bonnet-making, there was the bleaching to be performed; and it was by no means certain that this could be accomplished by means of an English sun, the difference between which and that of Italy or Connecticut was well known to be very great.

G.--My experiments have, I presume, completely removed this doubt. I think that the straw produced by me to the Society, and also some of the pieces of plat, are of a colour which no straw or plat can surpass. All that remains, therefore, is for me to give an account of the manner in which I cut and bleached the grass which I have submitted to the Society in the state of straw.

H.--First, as to the season of the year, all the straw, except that of one sort of couch-grass, and the long coppice-grass, which two were got in Sussex, were got from grass cut in Hertfordshire on the 21st of June. A grass head-land, in a wheat-field, had been mowed during the forepart of the day, and in the afternoon I went and took a handful here and a handful there out of the swaths. When I had collected as much as I could well carry, I took it to my friend's house, and proceeded to prepare it for bleaching, according to the information sent me from America by my son; that is to say, I put my grass into a shallow tub, put boiling water upon it until it was covered by the water, let it remain in that state for ten minutes, then took it out, and laid it very thinly on a closely-mowed lawn in a garden. But I should observe, that, before I put the grass into the tub, I tied it up in small bundles, or sheaves, each bundle being about six inches through at the butt-end. This was necessary, in order to be able to take the grass, at the end of ten minutes, out of the water, without throwing it into a confused mixture as to tops and tails. Being tied up in little bundles, I could easily, with a prong, take it out of the hot water. The bundles were put into a large wicker basket, carried to the lawn in the garden, and there taken out, one by one, and laid in swaths as before-mentioned.

I.--It was laid very thinly; almost might I say, that no stalk of grass covered another. The swaths were turned once a day. The bleaching was completed at the end of seven days from time of scalding and laying out. June is a fine month. The grass was, as it happened, cut on the longest day in the year; and the weather was remarkably fine and clear. But the grass which I afterwards cut in Sussex, was cut in the first week in August; and as to the weather my journal speaks thus:--

August, 1822.

2d.--Thunder and rain.--Began cutting grass. 3d.--Beautiful day. 4th.--Fine day. 5th.--Cloudy day--Began scalding grass, and laying it out.

6th.--Cloudy greater part of the day. 7th.--Same weather. 8th.--Cloudy and rather misty.--Finished cutting grass. 9th.--Dry but cloudy. 10th.--Very close and hot.--Packed up part of the grass. 11th, 12th, 13th, and

14th.--Same weather. 15th.--Hot and clear.--Finished packing the grass.

K.--The grass cut in Sussex was as well bleached as that cut in Hertfordshire; so that it is evident that we never can have a summer that will not afford sun sufficient for this business.

L.--The part of the straw used for platting; that part of the stalk which is above the upper joint; that part which is between the upper joint and the seed-branches. This part is taken out, and the rest of the straw thrown away. But the whole plant must be cut and bleached; because, if you were to take off, when green, the part above described, that part would wither up next to nothing. This part must die in company with the whole plants, and be separated from the other parts after the bleaching has been performed.

M.--The time of cutting must vary with the seasons, the situation, and the sort of grass. The grass which I got in Hertfordshire, than which nothing can, I think, be more beautiful, was, when cut, generally in bloom; just in bloom. The wheat was in full bloom; so that a good time for getting grass may be considered to be that when the wheat is in bloom. When I cut the grass in Sussex, the wheat was ripe, for reaping had begun; but that grass is of a very backward sort, and, besides, grew in the shade amongst coppice-wood and under trees, which stood pretty thick.

N.--As to the sorts of grass, I have to observe generally, that in proportion as the colour of the grass is deep; that is to say, getting further from the yellow, and nearer to the blue, it is of a deep and dead yellow when it becomes straw. Those kinds of grass are best which are, in point of colour, nearest to that of wheat, which is a fresh pale green. Another thing is, the quality of the straw as to pliancy and toughness. Experience must be our guide here. I had not time to make a large collection of sorts; but those which I have sent to you contain three sorts which are proved to be good. In my letter of the 3d instant I sent you sixteen pieces of plat and eight bunches of straw, having the seed heads on, in order to show the sorts of grass. The sixteenth piece of plat was American. The first piece was from wheat cut and bleached by me; the rest from grass cut and bleached by me. I will here, for fear of mistake, give a list of the names of the several sorts of grass, the straw of which was sent with my letter of the 3d instant, referring to the numbers, as placed on the plat and on the bunches of straw.

PIECES BUNCHES SORTS OF PLAT. OF STRAW. OF GRASS.

No 1.-- No. 1. --Wheat.

2.} { Melica Cærulea; or, Purple Melica 3.} 2 and 3 { Grass.

4.} { Agrostis Stolonifera; or, Fiorin Grass; 12.} 4 and 12 { that is to say, one sort of Couch-grass.

5.} 10.} 5 and 10 Lolium Perenne; or Ray-grass.

6.} { Avena Flavescens; or, Yellow Oat 11.} 6 and 11 { grass.

7.} { Cynosurus Cristatus; or Crested 15.} 7 and 15 { Dog's-tail grass.

8.} { Anthoxanthum Odoratum; or, Sweet 13.} 8 and 13 { scented Vernal grass.

9.} { Agrostis Canina; or, Brown Bent 14.} 9 and 14 { grass.

O.--These names are those given at the Botanical Garden at Kew. But the same English names are not in the country given to these sorts of grass. The Fiorin grass, the Yellow Oat-grass, and the Brown-Bent, are all called couch-grass; except that the latter is, in Sussex, called Red Robin. It is the native grass of the plains of Long Island; and they call it Red Top. The Ray-grass is the common field grass, which is, all over the kingdom, sown with clover. The farmers, in a great part of the kingdom, call it Bent, or Bennett, grass; and sometimes it is galled Darnel-grass. The Crested Dog's-tail goes, in Sussex, by the name of Hendonbent; for what reason I know not. The sweet-scented Vernal-grass I have never, amongst the farmers, heard any name for. Miss WOODHOUSE'S grass appears, from the plants that I saw in the Adelphi, to be one of the sorts of Couch-grass. Indeed, I am sure that it is a Couch-grass, if the plants I there saw came from her seed. My son, who went into Connecticut, who saw the grass growing, and who sent me home a specimen of it, is now in England: he was with me when I cut the grass in Sussex; and he says that Miss WOODHOUSE'S was a Couch-grass. However, it is impossible to look at the specimens of straw and of plat which I have sent you, without being convinced that there is no want of the raw material in England. I was, after my first hearing of the subject, very soon convinced that the grass grew in England; but I had great doubts as to the capacity of our sun. Those doubts my own experiments have completely removed; but then I was not aware of the great effect of the scalding, of which, by the way, Miss WOODHOUSE had said nothing, and the knowledge of which we owe entirely to my son James' journey into Connecticut.

P.--Having thus given you an account of the time and manner of cutting the grass, of the mode of cutting and bleaching; having given you the best account I am able, as to the sorts of grass to be employed in this business; and having, in my former communications, given you specimens of the plat wrought from the several sorts of straw, I might here close my letter; but as it may be useful to speak of the expense of cutting and bleaching, I shall trouble you with a few words relating to it. If there were a field of

Ray-grass, or of Crested Dog's-tail, or any other good sort, and nothing else growing with it, the expense of cutting would be very little indeed, seeing that the scythe or reap-hook would do the business at a great rate. Doubtless there will be such fields; but even if the grass have to be cut by the handful, my opinion is, that the expense of cutting and bleaching would not exceed fourpence for straw enough to make a large bonnet. I should be willing to contract to supply straw, at this rate, for half a million of bonnets. The scalding must constitute a considerable part of the expense; because there must be fresh water for every parcel of grass that you put in the tub.

When water has scalded one parcel of cold grass, it will not scald another parcel. Besides, the scalding draws out the sweet matter of the grass, and makes the water the colour of that horrible stuff called London porter. It would be very good, by-the-by, to give to pigs. Many people give hay-tea to pigs and calves; and this is grass-tea. To scald a large quantity, therefore would require means not usually at hand, and the scalding is an essential part of the business.

Perhaps, in a large and convenient farm-house, with a good brewing copper, good fuel and water handy, four or five women might scald a wagon load in a day; and a wagon would, I think, carry straw enough (in the rough) to furnish the means of making a thousand bonnets. However, the scalding might take place in the field itself, by means of a portable boiler, especially if water were at hand; and perhaps it would be better to carry the water to the field than to carry the grass to the farm-house, for there must be ground to lay it out upon the moment it has been scalded, and no ground can be so proper as the newly-mowed ground where the grass has stood. The space, too, must be large, for any considerable quantity of grass. As to all these things, however, the best and cheapest methods will soon be discovered when people set about the work with a view to profit.

Q.--The Society will want nothing from me, nor from any-body else, to convince it of the importance of this matter; but I cannot, in concluding these communications to you, Sir, refrain from making an observation or two on the consequences likely to arise out of these inquiries. The manufacture is alone of considerable magnitude. Not less than about five millions of persons in this kingdom have a dress which consists partly of manufactured straw; and a large part, and all the most expensive part, of the articles thus used, now come from abroad. In cases where you can get from abroad any article at less expense than you can get it at home, the wisdom of fabricating that article at home may be doubted. But, in this case, you get the raw material by labour performed at home, and the cost of that labour is not nearly so great as would be the cost of the mere carriage of the straw from a foreign country to this. If our own people had all plenty of employment, and that too more profitable to them and to the country than the turning of a part of our own grass into articles of dress, then it would be advisable still to import Leghorn bonnets; but the facts being the reverse, it is clear, that whatever money, or money's worth things, be sent out of the country, in exchange for Leghorn bonnets, is, while we have the raw material here for next to nothing, just so much thrown away. The Italians, it may be said, take some of our manufactures in exchange; and let us suppose, for the purpose of illustration, that they take cloth from Yorkshire. Stop the exchange between Leghorn and Yorkshire, and, does Yorkshire lose part of its custom? No: for though those who make the bonnets out of English grass, prevent the Leghorners from buying Yorkshire cloth, they, with the money which they now get, instead of its being got by the Leghorners, buy the Yorkshire cloth themselves; and they wear this cloth too, instead of its being worn by the people of Italy; ay, Sir, and many, now in rags, will be well clad, if the laudable object of the Society be effected.

Besides this, however, why should we not export the articles of this manufacture? To America we certainly should; and I should not be at all surprised if we were to export them to Leghorn itself.

R.--Notwithstanding all this, however, if the manufacture were of a description to require, in order to give it success, the collecting of the manufacturers together in great numbers, I should, however great the wealth that it might promise, never have done any thing to promote its establishment. The contrary is happily the case: here all is not only performed by hand, but by hand singly, without any combination of hands. Here there is no power of machinery or of chemistry wanted. All is performed out in the open fields, or sitting in the cottage. There wants no coal mines and no rivers to assist; no water-powers nor powers of fire. No part of the kingdom is unfit for the business. Every-where there are grass, water, sun, and women and children's fingers; and these are all that are wanted. But, the great thing of all is this; that, to obtain the materials for the making of this article of dress, at once so gay, so useful, and in some cases so expensive, there requires not a penny of capital. Many of the labourers now make their own straw hats to wear in summer. Poor rotten things, made out of straw of ripened grain. With what satisfaction will they learn that straw, twenty times as durable, to say nothing of the beauty, is to be got from every hedge? In short when the people are well and clearly informed of the facts, which I have through you, Sir, had the honour to lay before the Society, it is next to impossible that the manufacture should not become general throughout the country. In every labourer's house a pot of water can be boiled. What labourer's wife cannot, in the summer months, find time to cut and bleach grass enough to give her and her children work for a part of the winter? There is no necessity for all to be platters. Some may cut and bleach only. Others may prepare the straw, as mentioned in paragraph L. of this letter. And doubtless, as the farmers in Hertfordshire now sell their straw to the platters, grass collectors and bleachers and preparers would do the same. So that there is scarcely any country labourer's family that might not derive some advantage from this discovery; and, while I am convinced that this consideration has been by no means over-looked by the Society, it has been, I assure you, the great consideration of all with, Sir, your most obedient and most humble Servant, WM. COBBETT.

223. In the last edition, this closing part of the work, relative to the straw plat, was not presented to the public as a thing which admitted of no alteration; but, on the contrary, it was presented to the public with the following concluding remark: "In conclusion I have to observe, that I by no means send forth this essay as containing opinions and instructions that are to undergo no alteration. I am, indeed, endeavouring to teach others; but I am myself only a learner. Experience will, doubtless, make me much more perfect in a knowledge of the several parts of the subject; and the fruit of this experience I shall be careful to communicate to the public." I now proceed to make good this promise. Experience has proved that very beautiful and very fine plat can be made of the straw of divers kinds of grass. But the most ample experience has also proved to us that it is to the straw of wheat, that we are to look for a manufacture to supplant the Leghorn. This was mentioned as a strong suspicion in my former edition of this work. And I urged my readers to sow wheat for the purpose. The fact is now proved beyond all contradiction, that the straw of wheat or rye, but particularly of wheat, is the straw for this purpose. Finer plat may be made from the straw of grass than can possibly be made from the straw of wheat or rye: but the grass plat is, all of it, more or less brittle; and none of it has the beautiful and uniform colour of the straw of wheat. Since the last edition of this work, I have received packets of the straw from Tuscany, all of wheat; and, indeed, I am convinced that no other straw is any-thing like so well calculated for the purpose. Wheat straw bleaches better than any other. It has that fine, pale, golden colour which no other straw has; it is much more simple, more pliant than any other straw; and, in short, this is the material. I did not urge in vain. A good quantity of wheat was sowed for this purpose. A great deal of it has been well harvested; and I have the pleasure to know that several hundreds of persons are now employed in the platting of straw. One more year; one more crop of wheat; and another Leghorn bonnet will never be imported in England. Some great errors have been committed in the sowing of the wheat, and in the cutting of it. I shall now, therefore, availing myself of the experience which I have gained, offer to the public some observations on the sort of wheat to be sowed for this purpose; on the season for sowing; on the land to be used for the purpose; on the quantity of seed, and the

manner of sowing: on the season for cutting; on the manner of cutting, bleaching, and housing; on the platting; on the knitting, and on the pressing.

224. The SORT OF WHEAT. The Leghorn plat is all made of the straw of the spring wheat. This spring wheat is so called by us, because it is sowed in the spring, at the same time that barley is sowed. The botanical name of it is TRITICUM ÆSTIVUM. It is a small-grained bearded wheat. It has very fine straw; but experience has convinced me, that the little brown-grained winter wheat is just as good for the purpose. In short, any wheat will do. I have now in my possession specimens of plat made of both winter and spring wheat, and I see no difference at all. I am decidedly of opinion that the winter wheat is as good as the spring wheat for the purpose. I have plat, and I have straw both now before me, and the above is the result of my experience.

225. THE LAND PROPER FOR THE GROWING OF WHEAT. The object is to have the straw as small as we can get it. The land must not, therefore, be too rich; yet it ought not to be very poor. If it be, you get the straw of no length. I saw an acre this year, as beautiful as possible, sowed upon a light loam, which bore last year a fine crop of potatoes. The land ought to be perfectly clean, at any rate; so that, when the crop is taken off, the wheat straw may not be mixed with weeds and grass.

226. SEASON FOR SOWING. This will be more conveniently stated in paragraph 228.

227. QUANTITY OF SEED AND MANNER OF SOWING. When first this subject was started in 1821, I said, in the Register, that I would engage to grow as fine straw in England as the Italians could grow. I recommended then, as a first guess, fifteen bushels of wheat to the acre. Since that, reflection told me that that was not quite enough. I therefore recommended twenty bushels to the acre. Upon the beautiful acre which I have mentioned above, eighteen bushels, I am told, were sowed; fine and beautiful as it was, I think it would have been better if it had had twenty bushels; twenty bushels, therefore, is what I recommend. You must sow broad cast, of course, and you must take great pains to cover the seed well. It must be a good even-handed seedsman, and there must be very nice covering.

228. SEASON FOR CUTTING. Now, mind, it is fit to cut in just about one week after the bloom has dropped. If you examine the ear at that time, you will find the grain just beginning to be formed, and that is precisely the time to cut the wheat: The straw has then got its full substance in it. But I must now point out a very material thing. It is by no means desirable to have all your wheat fit to cut at the same time. It is a great misfortune, indeed, so to have it. If fit to cut altogether, it ought to be cut all at the same time; for supposing you to have an acre, it will require a fortnight or three weeks to cut it and bleach it, unless you have a very great number of hands, and very great vessels to prepare water in. Therefore, if I were to have an acre of wheat for this, purpose, and were to sow all spring wheat, I would sow a twelfth part of the acre every week from the first week in March to the last week in May. If I relied partly upon winter wheat, I would sow some every month, from the latter end of September to March. If I employed the two sorts of wheat, or indeed if I employed only the spring wheat, the TRITICUM ÆSTIVUM, I should have some wheat fit to cut in June, and some not fit to cut till September. I should be sure to have a fair chance as to the weather. And, in short, it would be next to impossible for me

to fail of securing a considerable part of my crop. I beg the reader's particular attention to the contents of this paragraph.

229. MANNER OF CUTTING THE WHEAT. It is cut by a little reap-hook, close to the ground as possible. It is then tied in little sheaves, with two pieces of string, one near the butt, and the other about half-way up. This little bundle or sheaf ought to be six inches through at the butt, and no more. It ought not to be tied too tightly, lest the scalding should not be perfect.

230. MANNER OF BLEACHING. The little sheaves mentioned in the last paragraph are carried to a brewing mash, vat, or other tub. You must not put them into the tub in too large a quantity, lest the water get chilled before it get to the bottom. Pour on scalding water till you cover the whole of the little sheaves, and let the water be a foot above the top sheaves. When the sheaves have remained thus a full quarter of an hour, take them out with a prong, lay them in a clothes-basket, or upon a hurdle, and carry them to the ground where the bleaching is to be finished. This should be, if possible, a piece of grass land, where the grass is very short. Take the sheaves, and lay some of them along in a row; untie them, and lay the straw along in that row as thin as it can possibly be laid. If it were possible, no one straw ought to have another lying upon it, or across it. If the sun be clear, it will require to lie twenty-four hours thus, then to be turned, and lie twenty-four hours on the other side. If the sun be not very clear, it must lie longer. But the numerous sowings which I have mentioned will afford you so many chances, so many opportunities of having fine weather, that the risk about weather would necessarily be very small. If wet weather should come, and if your straw remain out in it any length of time, it will be spoiled; but, according to the mode of sowing above pointed out, you really could stand very little chance of losing straw by bad weather. If you had some straw out bleaching, and the weather were to appear suddenly to be about to change, the quantity that you would have out would not be large enough to prevent you from putting it under cover, and keeping it there till the weather changed.

231. HOUSING THE STRAW. When your straw is nicely bleached, gather it up, and with the same string that you used to tie it when green, tie it up again into little sheaves. Put it by in some room where there is no damp, and where mice and rats are not suffered to inhabit. Here it is always ready for use, and it will keep, I dare say, four or five years very well.

232. THE PLATTING. This is now so well understood that nothing need be said about the manner of doing the work. But much might be said about the measures to be pursued by land-owners, by parish officers, by farmers, and more especially by gentlemen and ladies of sense, public spirit, and benevolence of disposition. The thing will be done; the manufacture will spread itself all over this kingdom; but the exertions of those whom I have here pointed out might hasten the period of its being brought to perfection. And I beg such gentlemen and ladies to reflect on the vast importance of such manufacture, which it is impossible to cause to produce any-thing but good. One of the great misfortunes of England at this day is, that the land has had taken away from it those employments for its women and children which were so necessary to the well-being of the agricultural labourer. The spinning, the carding, the reeling, the knitting; these have been all taken away from the land, and given to the Lords of the Loom, the haughty lords of bands of abject slaves. But let the landholder mark how the change has operated to produce his ruin. He must have the labouring MAN and the labouring BOY; but, alas! he cannot

have these, without having the man's wife, and the boy's mother, and little sisters and brothers. Even Nature herself says, that he shall have the wife and little children, or that he shall not have the man and the boy. But the Lords of the Loom, the crabbed-voiced, hard-favoured, hard-hearted, puffed-up, insolent, savage and bloody wretches of the North have, assisted by a blind and greedy Government, taken all the employment away from the agricultural women and children. This manufacture of Straw will form one little article of employment for these persons. It sets at defiance all the hatching and scheming of all the tyrannical wretches who cause the poor little creatures to die in their factories, heated to eighty-four degrees. There will need no inventions of WATT; none of your horse powers, nor water powers; no murdering of one set of wretches in the coal mines, to bring up the means of murdering another set of wretches in the factories, by the heat produced from those coals; none of these are wanted to carry on this manufactory. It wants no combination laws; none of the inventions of the hard-hearted wretches of the North.

233. THE KNITTING. Upon this subject, I have only to congratulate my readers that there are great numbers of English women who can now knit, plat together, better than those famous Jewesses of whom we were told.

234. THE PRESSING. Bonnets and hats are pressed after they are made. I am told that a proper press costs pretty nearly a hundred pounds; but, then, that it will do prodigious deal of business. I would recommend to our friends in the country to teach as many children as they can to make the plat. The plat will be knitted in London, and in other considerable towns, by persons to whom it will be sold. It appears to me, at least, that this will be the course that the thing will take. However, we must leave this to time; and here I conclude my observations upon a subject which is deeply interesting to myself, and which the public in general deem to be of great importance.

235. POSTSCRIPT on brewing.--I think it right to say here, that, ever since I published the instructions for brewing by copper and by wooden utensils, the beer at my own house has always been brewed precisely agreeable to the instructions contained in this book; and I have to add, that I never have had such good beer in my house in all my lifetime, as since I have followed that mode of brewing. My table-beer, as well as my ale, is always as clear as wine. I have had hundreds and hundreds of quarters of malt brewed into beer in my house. My people could always make it strong enough and sweet enough; but never, except by accident, could they make it CLEAR. Now I never have any that is not clear. And yet my utensils are all very small; and my brewers are sometimes one labouring man, and sometimes another. A man wants showing how to brew the first time. I should suppose that we use, in my house, about seven hundred gallons of beer every year, taking both sorts together; and I can positively assert, that there has not been one drop of bad beer, and indeed none which has not been most excellent, in my house, during the last two years, I think it is, since I began using the utensils, and in the manner named in this book.

ICE-HOUSES.

236. First begging the reader to read again paragraph 149, I proceed here, in compliance with numerous requests to that effect, to describe, as clearly as I can, the manner of constructing the sort of Ice-houses therein mentioned. In England, these receptacles of frozen water are,

generally, under ground, and always, if possible, under the shade of trees, the opinion being, that the main thing, if not the only thing, is to keep away the heat. The heat is to be kept away certainly; but moisture is the great enemy of Ice; and how is this to be kept away either under ground, or under the shade of trees? Abundant experience has proved, that no thickness of wall, that no cement of any kind, will effectually resist moisture. Drops will, at times, be seen hanging on the under side of an arch of any thickness, and made of any materials, if it have earth over it, and even when it has the floor of a house over it; and wherever the moisture enters, the ice will quickly melt.

237. Ice-houses should therefore be, in all their parts, as dry as possible: and they should be so constructed, and the ice so deposited in them, as to ensure the running away of the meltings as quickly as possible, whenever such meltings come. Any-thing in way of drains or gutters, is too slow in its effect; and therefore there must be something that will not suffer the water proceeding from any melting, to remain an instant.

238. In the first place, then, the ice-house should stand in a place quite open to the sun and air; for whoever has travelled even but a few miles (having eyes in his head) need not be told how long that part of a road from which the sun and wind are excluded by trees, or hedges, or by any-thing else, will remain wet, or at least damp, after the rest of the road is even in a state to send up dust.

239. The next thing is to protect the ice against wet, or damp, from beneath. It should, therefore, stand on some spot from which water would run in every direction; and if the natural ground presents no such spot, it is no very great job to make it.

240. Then come the materials of which the house is to consist. These, for the reasons before-mentioned, must not be bricks, stones, mortar, nor earth; for these are all affected by the atmosphere; they will become damp at certain times, and dampness is the great destroyer of ice. The materials are wood and straw. Wood will not do; for, though not liable to become damp, it imbibes heat fast enough; and, besides, it cannot be so put together as to shut out air sufficiently. Straw is wholly free from the quality of becoming damp, except from water actually put upon it; and it can, at the same time, be placed on a roof, and on sides, to such a degree of thickness as to exclude the air in a manner the most perfect. The ice-house ought, therefore, to be made of posts, plates, rafters, laths, and straw. The best form is the circular; and the house, when made, appears as I have endeavoured to describe it in Fig. 3 of the plate.

241. FIG. 1, a, is the centre of a circle, the diameter of which is ten feet, and at this centre you put up a post to stand fifteen feet above the level of the ground, which post ought to be about nine inches through at the bottom, and not a great deal smaller at the top. Great care must be taken that this post be perfectly perpendicular; for, if it be not, the whole building will be awry.

242. b b b are fifteen posts, nine feet high, and six inches through at the bottom, without much tapering towards the top. These posts stand about two feet apart, reckoning from centre of post to centre of post, which leaves between each two a space of eighteen inches, c c c c are fifty-four posts, five feet high, and five inches through at the bottom, without much tapering towards the top. These posts stand about two feet apart, from centre of post to centre of post,

which leaves between each two a space of nineteen inches. The space between these two rows of posts is four feet in width, and, as will be presently seen, is to contain a wall of straw.

243. e is a passage through this wall; d is the outside door of the passage; f is the inside door; and the inner circle, of which a is the centre, is the place in which the ice is to be deposited.

244. Well, then, we have now got the posts up; and, before we talk of the roof of the house, or of the bed for the ice, it will be best to speak about the making of the wall. It is to be made of straw, wheat-straw, or rye-straw, with no rubbish in it, and made very smooth by the hand as it is put in. You lay it in very closely and very smoothly, so that if the wall were cut across, as at g g, in FIG. 2 (which FIG. 2 represents the whole building cut down through the middle, omitting the centre post,) the ends of the straws would present a compact face as they do after a cut of a chaff-cutter. But there requires something to keep the straw from bulging out between the posts. Little stakes as big as your wrist will answer this purpose. Drive them into the ground, and fasten, at top, to the plates, of which I am now to speak. The plates are pieces of wood which go all round both the circles, and are nailed on upon the tops of the posts. Their main business is to receive and sustain the lower ends of the rafters, as at m m and n n in FIG. 2. But to the plates also the stakes just mentioned must be fastened at top. Thus, then, there will be this space of four feet wide, having, on each side of it, a row of posts and stakes, not more than about six inches from each other, to hold up, and to keep in its place, this wall of straw.

245. Next come the rafters, as from s to n, FIG. 2. Carpenters best know what is the number and what the size of the rafters; but from s to m there need be only about half as many as from m to n. However, carpenters know all about this. It is their every-day work. The roof is forty-five degrees pitch, as the carpenters call it. If it were even sharper, it would be none the worse. There will be about thirty ends of rafters to lodge on the plate, as at m; and these cannot all be fastened to the top of the centre-post rising up from a; but carpenters know how to manage this matter, so as to make all strong and safe. The plate which goes along on the tops of the row of posts, b b b, must, of course, be put on in a somewhat sloping form; otherwise there would be a sort of hip formed by the rafters. However, the thatch is to be so deep, that this may not be of much consequence. Before the thatching begins, there are laths to put upon the rafters. Thatchers know all about this, and all that you have to do is, to take care that the thatcher tie the straw on well. The best way, in a case of such deep thatch, is to have a strong man to tie for the thatcher.

246. The roof is now raftered, and it is to receive a thatch of clean, sound, and well-prepared wheat or rye straw, four feet thick, as at h h in FIG. 2.

247. The house having now got walls and roof, the next thing is to make the bed to receive the ice. This bed is the area of the circle of which a is the centre. You begin by laying on the ground round logs, eight inches through, or thereabouts, and placing them across the area, leaving spaces between them of about a foot. Then, crossways on them, poles about four inches through, placed at six inches apart. Then, crossways on them, other poles, about two inches through, placed at three inches apart. Then, crossways on them, rods as thick as your finger, placed at an inch apart. Then upon these, small, clean, dry, last-winter-cut twigs, to the thickness of about two inches; or, instead of these twigs, good, clean, strong heath, free from grass and moss, and from rubbish of all sorts.

248. This is the bed for the ice to lie on; and as you see, the top of the bed will be seventeen inches from the ground. The pressure of the ice may, perhaps, bring it to fourteen, or to thirteen. Upon this bed the ice is put, broken and pummelled, and beaten down together in the usual manner.

249. Having got the bed filled with ice, we have next to shut it safely up. As we have seen, there is a passage (e). Two feet wide is enough for this passage; and, being as long as the wall is thick, it is of course, four feet long. The use of the passage is this: that you may have two doors, so that you may, in hot or damp weather, shut the outer door, while you have the inner door open. This inner door may be of hurdle-work, and straw, and covered, on one of the sides, with sheep-skins with the wool on, so as to keep out the external air. The outer-door, which must lock, must be of wood, made to shut very closely, and, besides, covered with skins like the other. At times of great danger from heat, or from wet, the whole of the passage may be filled with straw. The door (p. FIG. 3) should face the North, or between North and East.

250. As to the size of the ice-house, that must, of course, depend upon the quantity of ice that you may choose to have. A house on the above scale, is from w to x (FIG. 2) twenty-nine feet; from y to z (FIG. 2) nineteen feet. The area of the circle, of which a is the centre, is ten feet in diameter, and as this area contains seventy-five superficial feet, you will, if you put ice on the bed to the height of only five feet, (and you may put it on to the height of seven feet from the top of the bed,) you will have three hundred and seventy-five cubic feet of ice; and, observe, a cubic foot of ice will, when broken up, fill much more than a Winchester Bushel: what it may do as to an "IMPERIAL BUSHEL," engendered like Greek Loan Commissioners, by the unnatural heat of "PROSPERITY," God only knows! However, I do suppose, that, without making any allowance for the "cold fit," as Dr. Baring calls it, into which "late panic" has brought us; I do suppose, that even the scorching, the burning dog-star of "IMPERIAL PROSPERITY;" nay, that even DIVES himself, would hardly call for more than two bushels of ice in a day; for more than two bushels a day it would be, unless it were used in cold as well as in hot weather.

251. As to the expense of such a house, it could, in the country, not be much. None of the posts, except the main or centre-post, need be very straight. The other posts might be easily culled from tree-lops, destined for fire-wood. The straw would make all straight. The plates must of necessity be short pieces of wood; and, as to the stakes, the laths, and the logs, poles, rods, twigs, and heath, they would not all cost twenty shillings. The straw is the principal article; and, in most places, even that would not cost more than two or three pounds. If it last many years, the price could not be an object; and if but a little while, it would still be nearly as good for litter as it was before it was applied to this purpose. How often the bottom of the straw walls might want renewing I cannot say, but I know that the roof would with few and small repairs, last well for ten years.

252. I have said that the interior row of posts is to be nine feet high, and the exterior row five feet high. I, in each case, mean, with the plate inclusive. I have only to add, that by way of superabundant precaution against bottom wet, it will be well to make a sort of gutter, to receive the drip from the roof, and to carry it away as soon as it falls.

253. Now, after expressing a hope that I shall have made myself clearly understood by every reader, it is necessary that I remind him, that I do not pretend to pledge myself for the

complete success, nor for any success at all, of this mode of making ice-houses. But, at the same time, I express my firm belief, that complete success would attend it; because it not only corresponds with what I have seen of such matters; but I had the details from a gentleman who had ample experience to guide him, and who was a man on whose word and judgment I placed a perfect reliance. He advised me to erect an ice-house; but not caring enough about fresh meat and fish in summer, or at least not setting them enough above "prime pork" to induce me to take any trouble to secure the former, I never built an ice-house. Thus, then, I only communicate that in which I believe; there is, however, in all cases, this comfort, that if the thing fail as an ice-house, it will serve all generations to come as a model for a pig-bed.

ADDITION.

Kensington, Nov. 14th, 1831.

MANGEL WURZEL.

254. This last summer, I have proved that, as keep for cows, MANGEL WURZEL is preferable to SWEDISH TURNIPS, whether as to quantity or quality. But there needs no other alteration in the book, than merely to read mangel wurzel wherever you find Swedish turnip; the time of sowing, the mode and time of transplanting, the distances, and the cultivation, all being the same; and the only difference being in the application of the leaves, and in the time of harvesting the roots.

255. The leaves of the MANGEL WURZEL are of great value, especially in dry summers. You begin, about the third week in August, to take off by a downward pull, the leaves of the plants; and they are excellent food for pigs and cows; only observe this, that, if given to cows, there must be, for each cow, six pounds of hay a day, which is not necessary in the case of the Swedish turnips. These leaves last till the crop is taken up, which ought to be in the first week of November. The taking off of the leaves does good to the plants: new leaves succeed higher up; and the plant becomes longer than it otherwise would be, and, of course, heavier. But, in taking off the leaves, you must not approach too near to the top.

256. When you take the plants up in November, you must cut off the crowns and the remaining leaves; and they, again, are for cows and pigs. Then you put the roots into some place to keep them from the frost; and, if you have no place under cover, put them in pies, in the same manner as directed for the Swedish turnips. The roots will average in weight 10 lbs. each. They may be given to cows whole, or to pigs either, and they are better than the Swedish turnip for both animals; and they do not give any bad or strong taste to the milk and butter. But, besides this use of the mangel wurzel, there is another, with regard to pigs at least, of very great importance. The juice of this plant has so much of sweetness in it, that, in France, they make sugar of it; and have used the sugar, and found it equal in goodness to West India sugar. Many persons in England make beer of this juice, and I have drunk of this beer, and found it very good. In short, the juice is most excellent for the mixing of moist food for pigs. I am now (20th Nov. 1831) boiling it for this purpose. My copper holds seven strike-bushels; I put in three bushels of mangel wurzel cut into pieces two inches thick, and then fill the copper with water. I draw off as much of the liquor as I want to wet pollard, or meal, for little pigs or fatting-pigs,

and the rest, roots and all, I feed the yard-hogs with; and this I shall follow on till about the middle of May.

257. If you give boiled, or steamed, potatoes to pigs, there wants some liquor to mix with the potatoes; for the water in which potatoes have been boiled is hurtful to any animal that drinks it. But mix the potatoes with juice of mangel wurzel, and they make very good food for hogs of all ages. The mangel wurzel produces a larger crop than the Swedish turnip.

COBBETT'S CORN.

258. IF you prefer bread and pudding to milk, butter, and meat, this corn will produce, on your forty rods, forty bushels, each weighing 60 lbs. at the least; and more flour, in proportion, than the best white wheat. To make bread with it you must use two-thirds wheaten, or rye, flour; but in puddings this is not necessary. The puddings at my house are all made with this flour, except meat and fruit pudding; for the corn flour is not adhesive or clinging enough to make paste, or crust. This corn is the very best for hog-fatting in the whole world. I, last April, sent parcels of the seed into several counties, to be given away to working men: and I sent them instructions for the cultivation, which I shall repeat here.

259. I will first describe this corn to you. It is that which is sometimes called Indian corn; and sometimes people call it Indian wheat. It is that sort of corn which the disciples ate as they were going up to Jerusalem on the Sabbath-day. They gathered it in the fields as they went along and ate it green, they being "an hungered," for which you know they were reproved by the pharisees. I have written a treatise on this corn in a book which I sell for four shillings, giving a minute account of the qualities, the culture, the harvesting, and the various uses of this corn; but I shall here confine myself to what is necessary for a labourer to know about it, so that he may be induced to raise and may be enabled to raise enough of it in his garden to fat a pig of ten score.

260. There are a great many sorts of this corn. They all come from countries which are hotter than England. This sort, which my eldest son brought into England, is a dwarf kind, and is the only kind that I have known to ripen in this country: and I know that it will ripen in this country in any summer; for I had a large field of it in 1828 and 1829; and last year (my lease at my farm being out at Michaelmas, and this corn not ripening till late in October) I had about two acres in my garden at Kensington. Within the memory of man there have not been three summers so cold as the last, one after another; and no one so cold as the last. Yet my corn ripened perfectly well, and this you will be satisfied of if you be amongst the men to whom this corn is given from me. You will see that it is in the shape of the cone of a spruce fir; you will see that the grains are fixed round a stalk which is called the cob. These stalks or ears come out of the side of the plant, which has leaves like a flag, which plant grows to about three feet high, and has two or three and sometimes more, of these ears or bunches of grain. Out of the top of the plant comes the tassel, which resembles the plumes of feathers upon a hearse; and this is the flower of the plant.

261. The grain is, as you will see, about the size of a large pea, and there are from two to three hundred of these grains upon the ear, or cob. In my treatise, I have shown that, in America, all the hogs and pigs, all the poultry of every sort, the greater part of the oxen, and a

considerable part of the sheep, are fatted upon this corn; that it is the best food for horses; and that, when ground and dressed in various ways, it is used in bread, in puddings, in several other ways in families; and that, in short, it is the real staff of life, in all the countries where it is in common culture, and where the climate is hot. When used for poultry, the grain is rubbed off the cob. Horses, sheep, and pigs, bite the grain off, and leave the cob; but horned cattle eat cob and all.

262. I am to speak of it to you, however, only as a thing to make you some bacon, for which use it surpasses all other grain whatsoever. When the grain is in the whole ear, it is called corn in the ear; when it is rubbed off the cob, it is called shelled corn. Now, observe, ten bushels of shelled corn are equal, in the fatting of a pig, to fifteen bushels of barley; and fifteen bushels of barley, if properly ground and managed, will make a pig of ten score, if he be not too poor when you begin to fat him. Observe that everybody who has been in America knows, that the finest hogs in the world are fatted in that country; and no man ever saw a hog fatted in that country in any other way than tossing the ears of corn over to him in the sty, leaving him to bite it off the ear, and deal with it according to his pleasure. The finest and solidest bacon in the world is produced in this way.

263. Now, then, I know, that a bushel of shelled corn may be grown upon one single rod of ground sixteen feet and a half each way; I have grown more than that this last summer; and any of you may do the same if you will strictly follow the instructions which I am now about to give you.

1. Late in March (I am doing it now,) or in the first fortnight of April, dig your ground up very deep, and let it lie rough till between the seventh and fifteenth of May.

2. Then (in dry weather if possible) dig up the ground again, and make it smooth at top. Draw drills with a line two feet apart, just as you do drills for peas; rub the grains off the cob; put a little very rotten and fine manure along the bottom of the drill; lay the grains along upon that six inches apart; cover the grain over with fine earth, so that there be about an inch and a half on the top of the grain; pat the earth down a little with the back of a hoe to make it lie solid on the grain.

3. If there be any danger of slugs, you must kill them before the corn comes up if possible: and the best way to do this is to put a little hot lime in a bag, and go very early in the morning, and shake the bag all round the edges of the ground and over the ground. Doing this three or four times very early in a dewy morning, or just after a shower, will destroy all the slugs; and this ought to be done for all other crops as well as for that of corn.

4. When the corn comes up, you must take care to keep all birds off till it is two or three inches high; for the spear is so sweet, that the birds of all sorts are very apt to peck it off, particularly the doves and the larks and pigeons. As soon as it is fairly above ground, give the whole of the ground (in dry weather) a flat hoeing, and be sure to move all the ground close round the plants. When the weeds begin to appear again, give the ground another hoeing, but always in dry weather. When the plants get to be about a foot high or a little more, dig the ground between the rows, and work the earth up a little against the stems of the plants.

5. About the middle of August you will see the tassel springing up out of the middle of the plant, and the ears coming out of the sides. If weeds appear in the ground, hoe it again to kill the weeds, so that the ground may be always kept clean. About the middle of September you will find the grains of the ears to be full of milk, just in the state that the ears were at Jerusalem when the disciples cropped them to eat. From this milky state, they, like the grains of wheat, grow hard; and as soon as the grains begin to be hard, you should cut off the tops of the corn and the long flaggy leaves, and leave the ears to ripen upon the stalk or stem. If it be a warm summer, they will be fit to harvest by the last of October; but it does not signify if they remain out until the middle of November or even later. The longer they stay out, the harder the grain will be.

6. Each ear is covered in a very curious manner with a husk. The best way for you will be, when you gather in your crop to strip off the husks, to tie the ears in bunches of six or eight or ten, and to hang them up to nails in the walls, or against the beams of your house; for there is so much moisture in the cob that the ears are apt to heat if put together in great parcels. The room in which I write in London is now hung all round with bunches of this corn. The bunches may be hung up in a shed or stable for a while, and, when perfectly dry, they may be put into bags.

7. Now, as to the mode of using the corn; if for poultry, you must rub the grains off the cob; but if for pigs, give them the whole ears. You will find some of the ears in which the grain is still soft. Give these to your pig first; and keep the hardest to the last. You will soon see how much the pig will require in a day, because pigs, more decent than many rich men, never eat any more than is necessary to them. You will thus have a pig; you will have two flitches of bacon, two pig's cheeks, one set of souse, two griskins, two spare-ribs, from both which I trust in God you will keep the jaws of the Methodist parson; and if, while you are drinking a mug of your own ale, after having dined upon one of these, you drink my health, you may be sure that it will give you more merit in the sight of God as well as of man, than you would acquire by groaning the soul out of your body in responses to the blasphemous cant of the sleekheaded Methodist thief that would persuade you to live upon potatoes.

264. You must be quite sensible that I cannot have any motive but your good in giving you this advice, other than the delight which I take and the pleasure which I derive from doing that good. You are all personally unknown to me: in all human probability not one man in a thousand will ever see me. You have no more power to show your gratitude to me than you have to cause me to live for a hundred years. I do not desire that you should deem this a favour received from me. The thing is worth your trying, at any rate.

265. The corn is off by the middle of November. The ground should then be well manured, and deeply dug, and planted with EARLY YORK, or EARLY DWARF CABBAGES, which will be loaved in the latter end of April, and may be either sold or given to pigs, or cows, before the time to plant the corn again. Thus you have two very large crops on the same ground in the same year.

INDEX. PARAGRAPH

Agur 18

Bees 160

Bread, making of 77

Brewing Beer 20, 108 See also "POSTSCRIPT." Brewing-machine 41

Brougham, Mr. 41 Candles and Rushes 199

Castlereagh's and Mackintosh's Oratory 152 Combination Laws 108

Corn, Cobbett 258

Cows, keeping 111

Cusar, Mr. 86

Custom Laws 108

Drennen, Dr. 80

Dress, Household Goods, and Fuel 199 Ducks 169

Economy, meaning of the term 2, 3 Education 11

Ellman, Mr. 20, 60

Excise Laws 108

Fowls 176

Geese 167

Goats and Ewes 189 Hanning, Mr. Wm. 99 Hill, Mr. 98

Hops 202

Ice-houses 236

Leghorn 212

Libel Laws 108

Malthus, Parson 141

Mangel Wurzel 254

Mustard 198

Parks, Mr. 98

Paul, Saint 148

Peel's flimsy Dresses 152 Pigeons 181

Pigs, keeping 139 Pitt's false Money 152

Plat, English Straw 208 Porter, how to make 71 Potatoes 77

Rabbits 184

Salting Mutton and Beef 157 Stanhope, Lord 144

Swedish Turnips 207

Turkeys 171

Walter's and Stoddart's Paragraphs 152 Walter Scott's Poems 152

Want, the Parent of Crime 18

Wakefield, Mr. Edward 78, 99

Wilberforce's Potatoe-Diet 152

Winchelsea, Lord. 144

Woodhouse, Miss 213

Yeast 203

COBBETT'S POOR MAN'S FRIEND;

A DEFENCE OF THE RIGHTS OF THOSE WHO DO THE WORK, AND FIGHT THE BATTLES.

COBBETT'S POOR MAN'S FRIEND. NUMBER I.

TO THE WORKING CLASSES OF PRESTON.

Burghclere, Hampshire, 22d August, 1826.

MY EXCELLENT FRIENDS,

1. Amongst all the new, the strange, the unnatural, the monstrous things that mark the present times, or, rather, that have grown out of the present system of governing this country, there is, in my opinion, hardly any thing more monstrous, or even so monstrous, as the language that is now become fashionable, relative to the condition and the treatment of that part of the community which are usually denominated the POOR; by which word I mean to designate the persons who, from age, infirmity, helplessness, or from want of the means of gaining anything by labour, become destitute of a sufficiency of food or of raiment, and are in danger of perishing if they be not relieved. Such are the persons that we mean when we talk of THE POOR; and, I repeat, that amongst all the monstrous things of these monstrous days, nothing is, in my opinion,

so monstrous as the language which we now constantly hear relative to the condition and treatment of this part of the community.

2. Nothing can be more common than to read, in the newspapers, descriptions the most horrible of the sufferings of the Poor, in various parts of England, but particularly in the North. It is related of them, that they eat horse-flesh, grains, and have been detected in eating out of pig-troughs. In short, they are represented as being far worse fed and worse lodged than the greater part of the pigs. These statements of the newspapers may be false, or, at least, only partially true; but, at a public meeting of rate-payers, at Manchester, on the 17th of August, Mr. BAXTER, the Chairman, said, that some of the POOR had been starved to death, and that tens of thousands were upon the point of starving; and, at the same meeting, Mr. POTTER gave a detail, which showed that Mr. BAXTER'S general description was true. Other accounts, very nearly official, and, at any rate, being of unquestionable authenticity, concur so fully with the statements made at the Manchester Meeting, that it is impossible not to believe, that a great number of thousands of persons are now on the point of perishing for want of food, and that many have actually perished from that cause; and that this has taken place, and is taking place, IN ENGLAND.

3. There is, then, no doubt of the existence of the disgraceful and horrid facts; but that which is as horrid as are the facts themselves, and even more horrid than those facts, is the cool and unresentful language and manner in which the facts are usually spoken of. Those who write about the misery and starvation in Lancashire and Yorkshire, never appear to think that any body is to blame, even when the poor die with hunger. The Ministers ascribe the calamity to "over-trading;" the cotton and cloth and other master-manufacturers ascribe it to "a want of paper-money," or to the Corn-Bill; others ascribe the calamity to the taxes. These last are right; but what have these things to do with the treatment of the poor? What have these things to do with the horrid facts relative to the condition and starvation of English people? It is very true, that the enormous taxes which we pay on account of loans made to carry on the late unjust wars, on account of a great standing army in time of peace, on account of pensions, sinecures and grants, and on account of a Church, which, besides, swallows up so large a part of the produce of the land and the labour; it is very true, that these enormous taxes, co-operating with the paper-money and its innumerable monopolies; it is very true, that these enormous taxes, thus associated, have produced the ruin in trade, manufactures and commerce, and have, of course, produced the low wages and the want of employment; this is very true; but it is not less true, that, be wages or employment as they may, the poor are not to perish with hunger, or with cold, while the rest of the community have food and raiment more than the latter want for their own sustenance. The LAW OF ENGLAND says, that there shall be no person to suffer from want of food and raiment. It has placed officers in every parish to see that no person suffer from this sort of want; and lest these officers should not do their duty, it commands all the magistrates to hear the complaints of the poor, and to compel the officers to do their duty. The LAW OF ENGLAND has provided ample means of relief for the poor; for, it has authorized the officers, or overseers, to get from the rich inhabitants of the parish as much money as is wanted for the purpose, without any limit as to amount; and, in order that the overseers may have no excuse of inability to make people pay, the law has armed them with powers of a nature the most efficacious and the most efficient and most prompt in their operation. In short, the language of the LAW, to the overseer, is this: "Take care that no person suffer from hunger, or from cold; and that you may

be sure not to fail of the means of obeying this my command, I give you, as far as shall be necessary for this purpose, full power over all the lands, all the houses, all the goods, and all the cattle, in your parish." To the Justices of the Peace the LAW says: "Lest the overseer should neglect his duty; lest, in spite of my command to him, any one should suffer from hunger or cold, I command you to be ready to hear the complaint of every sufferer from such neglect; I command you to summon the offending overseer, and to compel him to do his duty."

4. Such being the language of the LAW, is it not a monstrous state of things, when we hear it commonly and coolly stated, that many thousands of persons in England are upon the point of starvation; that thousands will die of hunger and cold next winter; that many have already died of hunger; and when we hear all this, unaccompanied with one word of complaint against any overseer, or any justice of the peace! Is not this state of things perfectly monstrous? A state of things in which it appears to be taken for granted, that the LAW is nothing, when it is intended to operate as a protection to the poor! Law is always law: if one part of the law may be, with impunity, set at defiance, why not another and every other part of the law? If the law which provides for the succour of the poor, for the preservation of their lives, may be, with impunity, set at defiance, why should there not be impunity for setting at defiance the law which provides for the security of the property and the lives of the rich? If you, in Lancashire, were to read, in an account of a meeting in Hampshire, that, here, the farmers and gentlemen were constantly and openly robbed; that the poor were daily breaking into their houses, and knocking their brains out; and that it was expected that great part of them would be killed very soon: if you, in Lancashire were to hear this said of the state of Hampshire, what would you say? Say! Why, you would say, to be sure, "Where is the LAW; where are the constables, the justices, the juries, the judges, the sheriffs, and the hangmen? Where can that Hampshire be? It, surely, never can be in Old England. It must be some savage country, where such enormities can be committed, and where even those, who talk and who lament the evils, never utter one word in the way of blame of the perpetrators." And if you were called upon to pay taxes, or to make subscriptions in money, to furnish the means of protection to the unfortunate rich people in Hampshire, would you not say, and with good reason, "No: what should we do this for? The people of Hampshire have the SAME LAW that we have; they are under the same Government; let them duly enforce that law; and then they will stand in no need of money from us to provide for their protection."

5. This is what common sense says would be your language in such a case; and does not common sense say, that the people of Hampshire, and of every other part of England, will thus think, when they are told of the sufferings, and the starvation, in Lancashire and Yorkshire! The report of the Manchester ley-payers, which took place on the 17th of August, reached me in a friend's house in this little village; and when another friend, who was present, read, in the speeches of Mr. BAXTER and Mr. POTTER, that tens of thousands of Lancashire people were on the point of starvation, and that many had already actually died from starvation; and when he perceived, that even those gentlemen uttered not a word of complaint against either overseer or justices of the peace, he exclaimed: "What! are there no poor-laws in Lancashire? Where, amidst all this starvation, is the overseer? Where is the justice of the peace? Surely that Lancashire can never be in England?"

6. The observations of this gentleman are those which occur to every man of sense; when he hears the horrid accounts of the sufferings in the manufacturing districts; for, though

we are all well aware, that the burden of the poor-rates presses, at this time, with peculiar weight on the land-owners and occupiers, and on owners and occupiers of other real property, in those districts, we are equally well aware, that those owners and occupiers have derived great benefits from that vast population that now presses upon them. There is land in the parish in which I am now writing, and belonging to the farm in the house of which I am, which land would not let for 20s. a statute acre; while land, not so good, would let, in any part of Lancashire, near to the manufactories, at 60s. or 80s. a statute acre. The same may be said with regard to houses. And, pray, are the owners and occupiers, who have gained so largely by the manufacturing works being near their lands and houses; are they, now, to complain, if the vicinage of these same works causes a charge of rates there, heavier than exists here? Are the owners and occupiers of Lancashire to enjoy an age of advantages from the labours of the spinners and the weavers; and are they, when a reverse comes, to bear none of the disadvantages? Are they to make no sacrifices, in order to save from perishing those industrious and ever-toiling creatures, by the labours of whom their land and houses have been augmented in value, three, five, or perhaps tenfold? None but the most unjust of mankind can answer these questions in the affirmative.

7. But as greediness is never at a loss for excuses for the hard-heartedness that it is always ready to practise, it is said, that the whole of the rents of the land and the houses would not suffice for the purpose; that is to say, that if the poor rates were to be made so high as to leave the tenant no means of paying rent, even then some of the poor must go without a sufficiency of food. I have no doubt that, in particular instances, this would be the case.

But for cases like this the LAW has amply provided; for, in every case of this sort, adjoining parishes may be made to assist the hard pressed parish; and if the pressure becomes severe on these adjoining parishes, those next adjoining them may be made to assist; and thus the call upon adjoining parishes maybe extended till it reach all over the county. So good, so benignant, so wise, so foreseeing, and so effectual, is this, the very best of all our good old laws! This law or rather code of laws, distinguishes England from all the other countries in the world, except the United States of America, where, while hundreds of other English statutes have been abolished, this law has always remained in full force, this great law of mercy and humanity, which says, that no human being that treads English ground shall perish for want of food and raiment. For such poor persons as are unable to work, the law provides food and clothing; and it commands that work shall be provided for such as are able to work, and cannot otherwise get employment. This law was passed more than two hundred years ago. Many attempts have been made to chip it away, and some have been made to destroy it altogether; but it still exists, and every man who does not wish to see general desolation take place, will do his best to cause it to be duly and conscientiously executed.

8. Having now, my friends of Preston, stated what the law is, and also the reasons for its honest enforcement in the particular case immediately before us, I will next endeavour to show you that it is founded in the law of nature, and that, were it not for the provisions of this law, people would, according to the opinions of the greatest lawyers, have a right to take food and raiment sufficient to preserve them from perishing; and that such taking would be neither felony nor larceny. This is a matter of the greatest importance; it is a most momentous question; for if it be settled in the affirmative--if it be settled that it is not felony, nor larceny, to take other men's goods without their assent, and even against their will, when such taking is absolutely

necessary to the preservation of life, how great, how imperative, is the duty of affording, if possible, that relief which will prevent such necessity! In other words, how imperative it is on all overseers and justices to obey the law with alacrity; and how weak are those persons who look to "grants" and "subscriptions," to supply the place of the execution of this, the most important of all the laws that constitute the basis of English society! And if this question be settled in the affirmative; if we find the most learned of lawyers and most wise of men, maintaining the affirmative of this proposition; if we find them maintaining, that it is neither felony nor larceny to take food, in case of extreme necessity, though without the assent, and even against the will of the owner, what are we to think of those (and they are not few in number nor weak in power) who, animated with the savage soul of the Scotch feelosophers, would wholly abolish the poor-laws, or, at least, render them of little effect, and thereby constantly keep thousands exposed to this dire necessity!

9. In order to do justice to this great subject; in order to treat it with perfect fairness, and in a manner becoming of me and of you, I must take the authorities on both sides. There are some great lawyers who have contended that the starving man is still guilty of felony or larceny, if he take food to satisfy his hunger; but there are a greater number of other, and still greater, lawyers, who maintain the contrary. The general doctrine of those who maintain the right to take, is founded on the law of nature; and it is a saying as old as the hills, a saying in every language in the world, that "self-preservation is the first law of nature." The law of nature teaches every creature to prefer the preservation of its own life to all other things. But, in order to have a fair view of the matter before us, we ought to inquire how it came to pass, that the laws were ever made to punish men as criminals, for taking the victuals, drink, or clothing, that they might stand in need of. We must recollect, then, that there was a time when no such laws existed; when men, like the wild animals in the fields, took what they were able to take, if they wanted it. In this state of things, all the land and all the produce belonged to all the people in common. Thus were men situated, when they lived under what is called the law of nature; when every one provided, as he could, for his self-preservation.

10. At length this state of things became changed: men entered into society; they made laws to restrain individuals from following, in certain cases, the dictates of their own will; they protected the weak against the strong; the laws secured men in possession of lands, houses, and goods, that were called THEIRS; the words MINE and THINE, which mean my own and thy own, were invented to designate what we now call a property in things. The law necessarily made it criminal in one man to take away, or to injure, the property of another man. It was, you will observe, even in this state of nature, always a crime to do certain things against our neighbour. To kill him, to wound him, to slander him, to expose him to suffer from the want of food or raiment, or shelter. These, and many others, were crimes in the eye of the law of nature; but, to take share of a man's victuals or clothing; to go and insist upon sharing a part of any of the good things that he happened to have in his possession, could be no crime, because there was no property in anything, except in man's body itself. Now, civil society was formed for the benefit of the whole. The whole gave up their natural rights, in order that every one might, for the future, enjoy his life in greater security. This civil society was intended to change the state of man for the better. Before this state of civil society, the starving, the hungry, the naked man, had a right to go and provide himself with necessaries wherever he could find them. There would be sure to be some such necessitous persons in a state of civil society. Therefore, when civil society was

established, it is impossible to believe that it had not in view some provision for these destitute persons. It would be monstrous to suppose the contrary. The contrary supposition would argue, that fraud was committed upon the mass of the people in forming this civil society; for, as the sparks fly upwards, so will there always be destitute persons to some extent or other, in every community, and such there are to now a considerable extent, even in the UNITED STATES OF AMERICA; therefore, the formation of the civil society must have been fraudulent or tyrannical upon any other supposition than that it made provision, in some way or other, for destitute persons; that is to say, for persons unable, from some cause or other, to provide for themselves the food and raiment sufficient to preserve them from perishing. Indeed, a provision for the destitute seems essential to the lawfulness of civil society; and this appears to have been the opinion of BLACKSTONE, when, in the first Book and first Chapter of his Commentaries on the Laws of England, he says, "the law not only regards life and member, and protects every man in the enjoyment of them, but also furnishes him with every thing necessary for their support. For there is no man so indigent or wretched, but he may demand a supply sufficient for all the necessaries of life from the more opulent part of the community, by means of the several statutes enacted for the relief of the poor; a humane provision dictated by the principles of society."

11. No man will contend, that the main body of the people in any country upon earth, and of course in England, would have consented to abandon the rights of nature; to give up their right to enjoy all things in common; no man will believe, that the main body of the people would ever have given their assent to the establishing of a state of things which should make all the lands, and all the trees, and all the goods and cattle of every sort, private property; which should have shut out a large part of the people from having such property, and which should, at the same time, not have provided the means of preventing those of them, who might fall into indigence, from being actually starved to death! It is impossible to believe this. Men never gave their assent to enter into society on terms like these. One part of the condition upon which men entered into society was, that care should be taken that no human being should perish from want. When they agreed to enter into that state of things, which would necessarily cause some men to be rich and some men to be poor; when they gave up that right, which God had given them, to live as well as they could, and to take the means wherever they found them, the condition clearly was, the "principle of society;" clearly was, as BLACKSTONE defines it, that the indigent and wretched should have a right to "demand from the rich a supply sufficient for all the necessities of life."

12. If the society did not take care to act upon this principle; if it neglected to secure the legal means, of preserving the life of the indigent and wretched; then the society itself, in so far as that wretched person was concerned, ceased to have a legal existence. It had, as far as related to him, forfeited its character of legality. It had no longer any claim to his submission to its laws. His rights of nature returned: as far as related to him, the law of Nature revived in all its force: that state of things in which all men enjoyed all things in common was revived with regard to him; and he took, and he had a right to take, food and raiment, or, as Blackstone expresses it, "a supply sufficient for all the necessities of life." For, if it be true, as laid down by this English lawyer, that the principles of society; if it be true, that the very principles, or foundations of society dictate, that the destitute person shall have a legal demand for a supply from the rich, sufficient for all the necessities of life; if this be true, and true it certainly is, it follows of course

that the principles, that is, the base, or foundation, of society, is subverted, is gone; and that society is, in fact, no longer what it was intended to be, when the indigent, when the person in a state of extreme necessity, cannot, at once, obtain from the rich such sufficient supply: in short, we need go no further than this passage of BLACKSTONE, to show, that civil society is subverted, and that there is, in fact, nothing legitimate in it, when the destitute and wretched have no certain and legal resource.

13. But this is so important a matter, and there have been such monstrous doctrines and projects put forth by MALTHUS, by the EDINBURGH REVIEWERS, by LAWYER SCARLETT, by LAWYER NOLAN, by STURGES BOURNE, and by an innumerable swarm of persons who have been giving before the House of Commons what they call "evidence:" there have been such monstrous doctrines and projects put forward by these and other persons; and there seems to be such a lurking desire to carry the hostility to the working classes still further, that I think it necessary in order to show, that these English poor-laws, which have been so much calumniated by so many greedy proprietors of land; I think it necessary to show, that these poor-laws are the things which men of property, above all others, ought to wish to see maintained, seeing that, according to the opinions of the greatest and the wisest of men, they must suffer most in consequence of the abolition of those laws; because, by the abolition of those laws, the right given by the laws of nature would revive, and the destitute would take, where they now simply demand (as BLACKSTONE expresses it) in the name of the law. There has been some difference of opinion, as to the question, whether it be theft or no theft; or, rather, whether it be a criminal act, or not a criminal act, for a person, in a case of extreme necessity from want of food, to take food without the assent and even against the will, of the owner. We have, amongst our great lawyers, SIR MATTHEW HALE and SIR WILLIAM BLACKSTONE, who contend (though as we shall see, with much feebleness, hesitation, and reservation,) that it is theft, notwithstanding the extremity of the want; but there are many, and much higher authorities, foreign as well as English, on the other side. Before, however, I proceed to the hearing of these authorities, let me take a short view of the origin of the poor laws in England; for that view will convince us, that, though the present law was passed but a little more than two hundred years ago, there had been something to effect the same purpose ever since England had been called England.

14. According to the Common Law of England, as recorded in the MIRROUR OF JUSTICES, a book which was written before the Norman Conquest; a book in as high reputation, as a law-book, as any one in England; according to this book, CHAPTER 1st, SECTION 3d, which treats of the "First constitutions made by the antient kings;" According to this work, provision was made for the sustenance of the poor. The words are these: "It was ordained, that the poor should be sustained by parsons, by rectors of the church, and by the parishioners, so that none of them die for want of sustenance." Several hundred years later, the canons of the church show, that when the church had become rich, it took upon itself the whole of the care and expense attending the relieving of the poor. These canons, in setting forth the manner in which the tithes should be disposed of, say, "Let the priests set apart the first share for the building and ornaments of the church; let them distribute the second to the poor and strangers, with their own hands, in mercy and humility; and let them reserve the third part for themselves." This passage is taken from the canons of ELFRIC, canon 24th. At a later period, when the tithes had, in some places, been appropriated to convents, acts of Parliament were

passed, compelling the impropriators to leave, in the hands of their vicar, a sufficiency for the maintenance of the poor. There were two or three acts of this sort passed, one particularly in the twelfth year of RICHARD the Second, chapter 7th. So that here we have the most ancient book on the Common Law; we have the canons of the church at a later period; we have acts of Parliament at a time when the power and glory of England were at their very highest point; we have all these to tell us, that in England, from the very time that the country took the name, there was always a legal and secure provision for the poor, so that no person, however aged, infirm, unfortunate, or destitute, should suffer from want.

15. But, my friends, a time came when the provision made by the Common Law, by the Canons of the Church, and by the Acts of the Parliament coming in aid of those canons; a time arrived, when all these were rendered null by what is called the PROTESTANT REFORMATION. This "Reformation," As it is called, swept away the convents, gave a large part of the tithes to greedy courtiers, put parsons with wives and children into the livings, and left the poor without any resource whatsoever. This terrible event, which deprived England of the last of her possessions on the continent of Europe, reduced the people of England to the most horrible misery; from the happiest and best fed and best clad people in the world, it made them the most miserable, the most wretched and ragged of creatures. At last it was seen that, in spite of the most horrible tyranny that ever was exercised in the world, in spite of the racks and the gibbets and the martial law of QUEEN ELIZABETH, those who had amassed to themselves the property out of which the poor had been formerly fed, were compelled to pass a law to raise money, by way of tax, for relieving the necessities of the poor. They had passed many acts before the FORTY-THIRD year of the reign of this Queen Elizabeth; but these acts were all found to be ineffectual, till, at last, in the forty-third year of the reign: of this tyrannical Queen, and in the year of our Lord 1601, that famous act was passed, which has been in force until this day; and which, as I said before, is still in force, notwithstanding all the various attempts of folly and cruelty to get rid of it.

16. Thus, then, the present poor-laws are no new thing. They are no gift to the working people. You hear the greedy landowners everlastingly complaining against this law of QUEEN ELIZABETH. They pretend that it was an unfortunate law. They affect to regard it as a great INNOVATION, seeing that no such law existed before; but, as I have shown, a better law existed before, having the same object in view. I have shown, that the "Reformation," as it is called, had swept away that which had been secured to the poor by the Common Law, by the Canons of the Church, and by ancient Acts of Parliament. There was nothing new, then, in the way of benevolence towards the people, in this celebrated Act of Parliament of the reign of QUEEN ELIZABETH; and the landowners would act wisely by holding their tongues upon the subject; or, if they be too noisy, one may look into their GRANTS, and see if we cannot find something THERE to keep out the present parochial assessments.

17. Having now seen the origin of the present poor-laws, and the justice of their due execution, let us return to those authorities of which I was speaking but now, and an examination into which will show the extreme danger of listening to those projectors who would abolish the poor-laws; that is to say, who would sweep away that provision which was established in the reign of QUEEN ELIZABETH, from a conviction that it was absolutely necessary to preserve the peace of the country and the lives of the people. I observed before that there has been some

difference of opinion amongst lawyers as to the question, whether it be, or be not, theft, to take without his consent and against his will, the victuals of another, in order to prevent the taker from starving. SIR MATTHEW HALE and SIR WILLIAM BLACKSTONE say that it is theft. I am now going to quote the several authorities on both sides, and it will be necessary for me to indicate the works which I quote from by the words, letters, and figures which are usually made use of in quoting from these works. Some part of what I shall quote will be in Latin: but I shall put nothing in that language of which I will not give you the translation. I beg you to read these quotations with the greatest attention; for you will find, at the end of your reading, that you have obtained great knowledge upon the subject, and knowledge, too, which will not soon depart from your minds.

18. I begin with SIR MATTHEW HALE, (a Chief Justice of the Court of King's Bench in the reign of Charles the Second,) who, in his PLEAS OF THE CROWN, CHAP. IX., has the following passage, which I put in distinct paragraphs, and mark A, B, and C.

19. A. "Some of the casuists, and particularly COVARRUVIUS, Tom. I. De furti et rapinæ restitutione, § 3, 4, p. 473; and GROTIUS, de jure belli, ac pacis; lib. II. cap. 2. § 6, tell us, that in case of extreme necessity, either of hunger or clothing, the civil distributions of property cease, and by a kind of tacit condition the first community doth return, and upon this those common assertions are grounded: 'Quicquid necessitas cogit, defendit.' [Whatever necessity calls for, it justifies.] 'Necessitas est lex temporis et loci.' [Necessity is the law of time and place.] 'In casu extremæ necessitates omnia sunt communia.' [In case of extreme necessity, all things are in common;] and, therefore, in such case theft is no theft, or at least not punishable as theft; and some even of our own lawyers have asserted the same; and very bad use hath been made of this concession by some of the Jesuitical casuists of France, who have thereupon advised apprentices and servants to rob their masters, where they have been indeed themselves in want of necessaries, of clothes or victuals; whereof, they tell them, they themselves are the competent judges; and by this means let loose, as much as they can, by their doctrine of probability, all the ligaments of property and civil society."

20. B. "I do, therefore, take it, that, where persons live under the same civil government, as here in England, that rule, at least by the laws of England, is false; and, therefore, if a person being under necessity for want of victuals, or clothes, shall, upon that account, clandestinely, and 'animo furandi,' [with intent to steal,] steal another man's goods, it is felony, and a crime, by the laws of England, punishable with death; although, the judge before whom the trial is, in this case (as in other cases of extremity) be by the laws of England intrusted with a power to reprieve the offender, before or after judgment, in order to the obtaining the King's mercy. For, 1st, Men's properties would be under a strange insecurity, being laid open to other men's necessities, whereof no man can possibly judge, but the party himself. And, 2nd, Because by the laws of this kingdom [here he refers to the 43 Eliz. cap. 2] sufficient provision is made for the supply of such necessities by collections for the poor, and by the power of the civil magistrate. Consonant hereunto seems to be the law even among the Jews; if we may believe the wisest of kings. Proverbs vi. 30, 31. 'Men do not despise a thief, if he steal to satisfy his soul when he is hungry, but if he be found, he shall restore seven-fold, he shall give all the substance of his house.' It is true, death among them was not the penalty of theft, yet his necessity gave him no exception from the ordinary punishment inflicted by their law upon that offence."

21. C. "Indeed this rule, 'in casu extremæ necessitatis omnia sunt communia,' does hold, in some measure, in some particular cases, where, by the tacit consent of nations, or of some particular countries or societies, it hath obtained. First, among the Jews, it was lawful in case of hunger to pull ears of standing corn, and eat, (Matt. xii. 1;) and for one to pass through a vineyard, or olive-yard, to gather and eat without carrying away.

Deut. xxiii. 24, 25. SECOND, By the Rhodian law, and the common-maritime custom, if the common provision for the ship's company fail, the master may, under certain temperaments, break open the private chests of the mariners or passengers, and make a distribution of that particular and private provision for the preservation of the ship's company." Vide CONSOLATO DEL MARE, cap. 256. LE CUSTOMES DE LA MERE, p. 77.

22. SIR WILLIAM BLACKSTONE agrees, in substance, with HALE; but he is, as we shall presently see, much more eager to establish his doctrine; and, we shall see besides, that he has not scrupled to be guilty of misquoting, and of very shamefully garbling, the Scripture, in order to establish his point. We shall find him flatly contradicting the laws of England; but, he might have spared the Holy Scriptures, which, however, he has not done.

23. To return to HALE, you see he is compelled to begin with acknowledging that there are great authorities against him; and he could not say that GROTIUS was not one of the most virtuous as well as one of the most learned of mankind. HALE does not know very well what to do with those old sayings about the justification which hard necessity gives: he does not know what to do with the maxim, that, "in case of extreme necessity all things are owned in common." He is exceedingly puzzled with these ancient authorities, and flies off into prattle rather than argument, and tells us a story about "jesuitical" casuists in France, who advised apprentices and servants to rob their masters, and that they thus "let loose the ligaments of property and civil society." I fancy that it would require a pretty large portion of that sort of faith which induced this Protestant judge to send witches and wizards to the gallows; a pretty large portion of this sort of faith, to make us believe, that the "casuists of France," who, doubtless, had servants of their own, would teach servants to rob their masters! In short, this prattle of the judge seems to have been nothing more than one of those Protestant effusions which were too much in fashion at the time when he wrote.

24. He begins his second paragraph, or paragraph B., by saying, that he "takes it" to be so and so; and then comes another qualified expression; he talks of civil government "as here in England." Then he says, that the rule of GROTIUS and others, against which he has been contending, "he takes to be false, at least," says he, "by the laws of England." After he has made all these qualifications, he then proceeds to say that such taking is theft; that it is felony; and it is a crime which the laws of England punish with death! But, as if stricken with remorse at putting the frightful words upon paper; as if feeling shame for the law and for England itself, he instantly begins to tell us, that the judge who presides at the trial is intrusted, "by the laws of England," with power to reprieve the offender, in order to the obtaining of the King's mercy! Thus he softens it down. He will have it to be LAW to put a man to death in such a case; but he is ashamed to leave his readers to believe, that an English judge and an English king WOULD OBEY THIS LAW!

25. Let us now hear the reasons which he gives for this which he pretends to be law. His first reason is, that there would be no security for property, if it were laid open to the necessities of the indigent, of which necessities no man but the takers themselves could be the judge. He talks of a "strange insecurity;" but, upon my word, no insecurity could be half so strange as this assertion of his own. BLACKSTONE has just the same argument. "Nobody," says he, "would be a judge of the wants of the taker, but the taker himself;" and BLACKSTONE, copying the very words of HALE, talks of the "strange insecurity" arising from this cause. Now, then, suppose a man to come into my house, and to take away a bit of bacon. Suppose me to pursue him and seize him. He would tell me that he was starving for want of food. I hope that the bare statement would induce me, or any man in the world that I do call or ever have called my friend, to let him go without further inquiry; but, if I chose to push the matter further, there would be the magistrate. If he chose to commit the man, would there not be a jury and a judge to receive evidence and to ascertain whether the extreme necessity existed or not?

26. Aye, says Judge HALE; but I have another reason, a devilish deal better than this, "and that is, the act of the 43d year of the reign of QUEEN ELIZABETH!" Aye, my old boy, that is a thumping reason! "Sufficient provision is made for the supply of such necessities by collections for the poor, and by the power of the civil magistrate." Aye, aye! that is the reason; and, Mr. SIR MATTHEW HALE, there is no other reason, say what you will about the matter. There stand the overseer and the civil magistrate to take care that such necessities be provided for; and if they did not stand there for that purpose, the law of nature would be revived in behalf of the suffering creature.

27. HALE, not content however with this act of QUEEN ELIZABETH, and still hankering after this hard doctrine, furbishes up a bit of Scripture, and calls Solomon the wisest of kings on account of these two verses which he has taken. HALE observes, indeed, that the Jews did not put thieves to death; but, to restore seven-fold was the ordinary punishment, inflicted by their law, for theft; and here, says he, we see, that the extreme necessity gave no exemption. This was a piece of such flagrant sophistry on the part of HALE, that he could not find in his heart to send it forth to the world without a qualifying observation; but even this qualifying observation left the sophistry still so shameful, that his editor, Mr. EMLYN, who published the work under authority of the House of Commons, did not think it consistent with his reputation to suffer this passage to go forth unaccompanied with the following remark: "But their (the Jews') ordinary punishment being entirely pecuniary, could affect him only when he was found in a condition to answer it; and therefore the same reasons which could justify that, can, by no means, be extended to a corporal, much less to a capital punishment." Certainly: and this is the fair interpretation of these two verses of the Proverbs. PUFFENDORF, one of the greatest authorities that the world knows anything of, observes, upon the argument built upon this text of Scripture, "It may be objected, that, in Proverbs, chap. vi. verses 30, 31, he is called a thief, and pronounced obnoxious to the penalty of theft, who steals to satisfy his hunger; but whoever closely views and considers that text will find that the thief there censured is neither in such extreme necessity as we are now supposing, nor seems to have fallen into his needy condition merely by ill fortune, without his own idleness or default: for the context implies, that he had a house and goods sufficient to make seven-fold restitution; which he might have either sold or pawned; a chapman or creditor being easily to be met with in times of plenty and peace;

for we have no grounds to think that the fact there mentioned is supposed to be committed, either in time of war, or upon account of the extraordinary price of provisions."

28. Besides this, I think it is clear that these two verses of the Proverbs do not apply to one and the same person; for in the first verse it is said, that men do not despise a thief if he steal to satisfy his soul when he is hungry. How, then, are we to reconcile this with morality? Are we not to despise a thief? It is clear that the word thief does not apply to the first case; but to the second case only; and that the distinction was here made for the express purpose of preventing the man who took food to relieve his hunger from being confounded with the thief. Upon any other interpretation, it makes the passage contain nonsense and immorality; and, indeed, GROTIUS says that the latter text does not apply to the person mentioned in the former. The latter text could not mean a man taking food from necessity. It is impossible that it can mean that; because the man who was starving for want of food could not have seven-fold; could not have any substance in his house. But what are we to think of JUDGE BLACKSTONE, who, in his Book IV., chap. 2, really garbles these texts of Scripture. He clearly saw the effect of the expression, "MEN DO NOT DESPISE;" he saw what an awkward figure these words made, coming before the words "A THIEF;" he saw that, with these words in the text, he could never succeed in making his readers believe that a man ought to be hanged for taking food to save his life. He clearly saw that he could not make men believe that God had said this, unless he could, somehow or other, get rid of those words about NOT DESPISING the thief that took victuals when he was hungry. Being, therefore, very much pestered and annoyed by these words about NOT DESPISING, what does he do but fairly leave them out! And not only leave them out, but leave out a part of both the verses, keeping in that part of each that suited him, and no more; nay, further, leaving out one word, and putting in another, giving a sense to the whole which he knew well never was intended. He states the passage to be this: "If a thief steal to satisfy his soul when he is hungry, he shall restore seven-fold, and shall give all the substance of his house." No broomstick that ever was handled would have been too heavy or too rough for the shoulders of this dirty-souled man.

HALE, with all his desire to make out a case in favour of severity, has given us the words fairly: but this shuffling fellow; this smooth-spoken and mean wretch, who is himself thief enough, God knows, if stealing other men's thoughts and words constitute theft; this intolerably mean reptile has, in the first place, left out the words "men do not despise:" then he has left out the words at the beginning of the next text, "but if he be found." Then in place of the "he," which comes before the words "shall give" he puts the word "and;" and thus he makes the whole apply to the poor creature that takes to satisfy his soul when he is hungry! He leaves out every mitigating word of the Scripture; and, in his reference, he represents the passage to be in one verse! Perhaps, even in the history of the conduct of crown-lawyers, there is not to be found mention of an act so coolly bloody-minded as this. It has often been said of this BLACKSTONE, that he not only lied himself, but made others lie; he has here made, as far as he was able, a liar of King Solomon himself: he has wilfully garbled the Holy Scripture; and that, too, for the manifest purpose of justifying cruelty in courts and judges; for the manifest purpose of justifying the most savage oppression of the poor.

29. After all, HALE has not the courage to send forth this doctrine of his, without allowing that the case of extreme necessity does, "in some measure," and "in particular cases,"

and, "by the tacit or silent consent of nations," hold good! What a crowd of qualifications is here! With what reluctance he confesses that which all the world knows to be true, that the disciples of JESUS CHRIST pulled off, without leave, the ears of standing corn, and ate them "being an hungered." And here are two things to observe upon. In the first place this corn was not what we call corn here in England, or else it would have been very droll sort of stuff to crop off and eat. It was what the Americans call Indian corn, what the French call Turkish corn; and what is called corn (as being far surpassing all other in excellence) in the Eastern countries where the Scriptures were written. About four or five ears of this corn, of which you strip all the husk off in a minute, are enough for a man's breakfast or dinner; and by about the middle of August this corn is just as wholesome and as efficient as bread. So that, this was something to take and eat without the owner's leave; it was something of value; and observe, that the Pharisees, though so strongly disposed to find fault with everything that was done by Jesus Christ and his disciples, did not find fault of their taking the corn to eat; did not call them thieves; did not propose to punish them for theft; but found fault of them only for having plucked the corn on the Sabbath-day! To pluck the corn was to do work, and these severe critics found fault of this working on the Sabbath-day. Then, out comes another fact, which HALE might have noticed if he had chosen it; namely, that our Saviour reminds the Pharisees that "DAVID and his companions, being an hungered, entered into the House of God, and did eat the show-bread, to eat which was unlawful in any-body but the priests." Thus, that which would have been sacrilege under any other circumstances; that which would have been one of the most horrible of crimes against the law of God, became no crime at all when committed by a person pressed by hunger.

30. Nor has JUDGE HALE fairly interpreted the two verses of DEUTERONOMY. He represents the matter thus: that, if you be passing through a vineyard or an olive-yard you may gather and eat, without being deemed a thief. This interpretation would make an Englishman believe that the Scripture allowed of this taking and eating, only where there was a lawful foot-way through the vineyard. This is a very gross misrepresentation of the matter; for if you look at the two texts, you will find, that they say that, "when thou comest into;" that is to say, when thou enterest or goest into, "thy neighbour's vineyard, then thou mayest eat grapes thy fill at thine own pleasure, but thou shalt not put any in thy vessel;" that is to say, that you should not go and make wine in his vineyard and carry it away. Then in case of the corn, precisely the same law is laid down. You may pluck with your hand; but not use the hook or a sickle. Nothing can be plainer than this: no distinction can be wiser, nor more just. HALE saw the force of it; and therefore, as these texts made very strongly against him, he does not give them at full length, but gives us a misrepresenting abbreviation.

31. He had, however, too much regard for his reputation to conclude without acknowledging the right of seizing on the provisions of others at sea. He allows that private chests may be broken open to prevent men from dying with hunger at sea. He does not stop to tell us why men's lives are more precious on sea than on land. He does not attempt to reconcile these liberties given by the Scripture, and by the maritime laws, with his own hard doctrine. In short, he brings us to this at last: that he will not acknowledge, that it is not theft to take another man's goods, without his consent, under any circumstances; but, while he will not acknowledge this, he plainly leaves us to conclude, that no English judge and no English king will ever punish a poor creature that takes victuals to save himself from perishing; and he plainly leaves us to conclude, that it is the poor-laws of England; that it is their existence and their due execution,

which deprive everybody in England of the right to take food and raiment in case of extreme necessity.

32. Here I agree with him most cordially; and it is because I agree with him in this, that I deprecate the abominable projects of those who would annihilate the poor-laws, seeing that it is those very poor-laws which give, under all circumstances, really legal security to property. Without them, cases must frequently arise, which would, according to the law of nature, according to the law of God, and as we shall see before we have done, according to the law of England, bring us into a state, or, at least, bring particular persons into a state, which as far as related to them, would cause the law of nature to revive, and to make all things to be owned in common. To adhere, then, to these poor-laws; to cause them to be duly executed, to prevent every encroachment upon them, to preserve them as the apple of our eye, are the duty of every Englishman, as far as he has capacity so to do.

33. I have, my friends, cited, as yet, authorities only on one side of this great subject, which it was my wish to discuss in this one Number. I find that to be impossible without leaving undone much more than half my work. I am extremely anxious to cause this matter to be well understood, not only by the working classes, but by the owners of the land and the magistrates. I deem it to be of the greatest possible importance; and, while writing on it, I address myself to you, because I most sincerely declare that I have a greater respect for you than for any other body of persons that I know any thing of. The next Number will conclude the discussion of the subject. The whole will lie in a very small compass. Sixpence only will be the cost of it. It will creep about, by degrees, over the whole of this kingdom. All the authorities, all the arguments, will be brought into this small compass; and I do flatter myself that many months will not pass over our heads, before all but misers and madmen will be ashamed to talk of abolishing the poor-rates and of supporting the needy by grants and subscriptions.

I am, Your faithful friend and Most obedient servant, WM. COBBETT. NUMBER II.

Bollitree Castle, Herefordshire, 22d Sept. 1826.

MY EXCELLENT FRIENDS,

34. In the last Number, paragraph 33, I told you, that I would, in the present Number, conclude the discussion of the great question of theft, or no theft, in a case of taking another's goods without his consent, or against his will, the taker being pressed by extreme necessity. I laid before you; in the last Number, JUDGE HALE'S doctrine upon the subject; and I there mentioned the foul conduct of BLACKSTONE, the author of the "Commentaries on the Laws of England." I will not treat this unprincipled lawyer, this shocking court sycophant; I will not treat him as he has treated King Solomon and the Holy Scriptures; I will not garble, misquote, and belie him, as he garbled, misquoted, and belied them; I will give the whole of the passage to which I allude, and which my readers may find in the Fourth Book of his Commentaries. I request you to read it with great attention; and to compare it, very carefully, with the passage that I have quoted from SIR MATTHEW HALE, which you will find in paragraphs from 19 to 21 inclusive. The passage from BLACKSTONE is as follows:

35. "There is yet another case of necessity, which has occasioned great speculation among the writers upon general law; viz., whether a man in extreme want of food or clothing may justify stealing either, to relieve his present necessities. And this both GROTIUS and PUFFENDORF, together with many other of the foreign jurists, hold in the affirmative; maintaining by many ingenious, humane, and plausible reasons, that in such cases the community of goods by a kind of tacit concession of society is revived.

And some even of our own lawyers have held the same; though it seems to be an unwarranted doctrine, borrowed from the notions of some civilians: at least it is now antiquated, the law of England admitting no such excuse at present. And this its doctrine is agreeable not only to the sentiments of many of the wisest ancients, particularly CICERO, who holds that 'suum cuique incommodum ferendum est, potius quam de alterius commodis detrahendum;' but also to the Jewish law, as certified by King Solomon himself: 'If a thief steal to satisfy his soul when he is hungry, he shall restore seven-fold, and shall give all the substance of his house:' which was the ordinary punishment for theft in that kingdom. And this is founded upon the highest reason: for men's properties would be under a strange insecurity, if liable to be invaded according to the wants of others; of which wants no man can possibly be an adequate judge, but the party himself who pleads them. In this country especially, there would be a peculiar impropriety in admitting so dubious an excuse; for by our laws such a sufficient provision is made for the poor by the power of the civil magistrate, that it is impossible that the most needy stranger should ever be reduced to the necessity of thieving to support nature. This case of a stranger is, by the way, the strongest instance put by Baron PUFFENDORF, and whereon he builds his principal arguments; which, however they may hold upon the continent, where the parsimonious industry of the natives orders every one to work or starve, yet must lose all their weight and efficacy in England, where charity is reduced to a system, and interwoven in our very constitution. Therefore, our laws ought by no means to be taxed with being unmerciful, for denying this privilege to the necessitous; especially when we consider, that the king, on the representation of his ministers of justice, hath a power to soften the law, and to extend mercy in cases of peculiar hardship. An advantage which is wanting in many states, particularly those which are democratical: and these have in its stead introduced and adopted, in the body of the law itself, a multitude of circumstances tending to alleviate its rigour. But the founders of our constitution thought it better to vest in the crown the power of pardoning peculiar objects of compassion, than to countenance and establish theft by one general undistinguishing law."

36. First of all, I beg you to observe, that this passage is merely a flagrant act of theft, committed upon JUDGE HALE; next, you perceive, that which I noticed in paragraph 28, a most base and impudent garbling of the Scriptures. Next, you see, that BLACKSTONE, like HALE, comes, at last, to the poor-laws; and tells us that to take other men's goods without leave, is theft, because "charity is here reduced to a system, and interwoven in our very constitution." That is to say, to relieve the necessitous; to prevent their suffering from want; completely to render starvation impossible, makes a part of our very constitution. "THEREFORE, our laws ought by no means to be taxed with being unmerciful for denying this privilege to the necessitous." Pray mark the word therefore. You see, our laws, he says, are not to be taxed with being unmerciful in deeming the necessitous taker a thief. And why are they not to be deemed unmerciful? BECAUSE the laws provide effectual relief for the necessitous. It follows, then, of course, even according to BLACKSTONE himself, that if the Constitution had not provided

this effectual relief for the necessitous, then the laws would have been unmerciful in deeming the necessitous taker a thief.

37. But now let us hear what that GROTIUS and that PUFFENDORF say; let us hear what these great writers on the law of nature and of nations say upon this subject. BLACKSTONE has mentioned the names of them both; but he has not thought proper to notice their arguments, much less has he attempted to answer them. They are two of the most celebrated men that ever wrote; and their writings are referred to as high authority, with regard to all the subjects of which they have treated. The following is a passage from GROTIUS, on War and Peace, Book II., chap. 2.

38. "Let us see, further, what common right there appertains to men in those things which have already become the property of individuals. Some persons, perchance, may consider it strange to question this, as proprietorship seems to have absorbed all that right which arose out of a state of things in common. But it is not so. For, it is to be considered, what was the intention of those who first introduced private property, which we may suppose to have been such, as to deviate as little as possible from natural equity. For if even written laws are to be construed in that sense, as far as it is practicable, much more so are customs, which are not fettered by the chains of writers.--Hence it follows, first, that, in case of extreme necessity, the pristine right of using things revives, as much as if they had remained in common; because, in all human laws, as well as in the law of private property, this case of extreme necessity appears to have been excepted.--So, if the means of sustenance, as in case of a sea-voyage, should chance to fail, that which any individual may have, should be shared in common. And thus, a fire having broken out, I am justified in destroying the house of my neighbour, in order to preserve my own house; and I may cut in two the ropes or cords amongst which any ship is driven, if it cannot be otherwise disentangled. All which exceptions are not made in the written law, but are presumed.--For the opinion has been acknowledged amongst Divines, that, if any one, in such case of necessity, take from another person what is requisite for the preservation of his life, he does not commit a theft. The meaning of which definition is not, as many contend, that the proprietor of the thing be bound to give to the needy upon the principle of charity; but, that all things distinctly vested in proprietors ought to be regarded as such with a certain benign acknowledgment of the primitive right. For if the original distributors of things were questioned, as to what they thought about this matter, they would reply what I have said.

Necessity, says Father SENECA, the great excuse for human weakness, breaks every law; that is to say, human law, or law made after the manner of man."

39. "But cautions ought to be had, for fear this license should be abused: of which the principal is, to try, in every way, whether the necessity can be avoided by any other means; for instance, by making application to the magistrate, or even by trying whether the use of the thing can, by entreaties, be obtained from the proprietor. PLATO permits water to be fetched from the well of a neighbour upon this condition alone, that the person asking for such permission shall dig in his own well in search of water as far as the chalk: and SOLON, that he shall dig in his own well as far as forty cubits. Upon which PLUTARCH adds, that he judged that necessity was to be relieved, not laziness to be encouraged."

40. Such is the doctrine of this celebrated civilian. Let us now hear PUFFENDORF; and you will please to bear in mind, that both these writers are of the greatest authority upon all subjects connected with the laws of nature and of nations. We read in their works the result of an age of study: they have been two of the great guides of mankind ever since they wrote: and, we are not to throw them aside, in order to listen exclusively to Parson HAY, to HULTON OF HULTON, or to NICHOLAS GRIMSHAW. They tell us what they, and what other wise men, deemed to be right; and, as we shall by and by see, the laws of England, so justly boasted of by our ancestors, hold precisely the same language with these celebrated men. After the following passage from PUFFENDORF, I shall show you what our own lawyers say upon the subject; but I request you to read the following passage with the greatest attention.

41. "Let us inquire, in the next place, whether the necessity of preserving our life can give us any right over other men's goods, so as to make it allowable for us to seize on them for our relief, either secretly, or by open force, against the owner's consent. For the more clear and solid determination of which point, we think it necessary to hint in short on the causes upon which distinct properties were first introduced in the world; designing to examine them more at large in their proper place. Now the main reasons on which properties are founded, we take to be these two; that the feuds and quarrels might be appeased which arose in the primitive communion of things, and that men might be put under a kind of necessity of being industrious, every one being to get his maintenance by his own application and labour. This division, therefore, of goods, was not made, that every person should sit idly brooding over the share of wealth he had got, without assisting or serving his fellows; but that any one might dispose of his things how he pleased; and if he thought fit to communicate them to others, he might, at least, be thus furnished with an opportunity of laying obligations on the rest of mankind. Hence, when properties were once established, men obtained a power, not only of exercising commerce to their mutual advantage and gain, but likewise of dispensing more largely in the works of humanity and beneficence; whence their diligence had procured them a greater share of goods than others: whereas before, when all things lay in common, men could lend one another no assistance but what was supplied by their corporeal ability, and could be charitable of nothing but of their strength. Further, such is the force of property, that the proprietor hath a right of delivering his goods with his own hands; even such as he is obliged to give to others. Whence it follows, that when one man has anything owing from another, he is not presently to seize on it at a venture, but ought to apply himself to the owner, desiring to receive it from his disposal. Yet in case the other party refuse thus to make good his obligation, the power and privilege of property doth not reach so far as that the things may not be taken away without the owner's consent, either by the authority of the magistrate in civil communities, or in a state of nature, by violence and hostile force. And though in regard to bare Natural Right, for a man to relieve another in extremity with his goods, for which he himself hath not so much occasion, be a duty obliging only imperfectly, and not in the manner of a debt, since it arises wholly from the virtue of humanity; yet there seems to be no reason why, by the additional force of a civil ordinance, it may not be turned into a strict and perfect obligation. And this Seldon observes to have been done among the Jews; who, upon a man's refusing to give such alms as were proper for him, could force him to it by an action at law. It is no wonder, therefore, that they should forbid their poor, on any account, to seize on the goods of others, enjoining them to take only what private persons, or the public officers, or stewards of alms, should give them on their petition. Whence the stealing of what was another's, though upon extreme necessity, passed in that state for theft or rapine. But now

supposing under another government the like good provision is not made for persons in want, supposing likewise that the covetous temper of men of substance cannot be prevailed on to give relief, and that the needy creature is not able, either by his work or service, or by making sale of anything that he possesses, to assist his present necessity, must he, therefore, perish with famine? Or can any human institution bind me with such a force that, in case another man neglects his duty towards me, I must rather die, than recede a little from the ordinary and regular way of acting? We conceive, therefore, that such a person doth not contract the guilt of theft, who happening, not through his own fault, to be in extreme want, either of necessary food, or of clothes to preserve him from the violence of the weather, and cannot obtain them from the voluntary gift of the rich, either by urgent entreaties, or by offering somewhat equivalent in price, or by engaging to work it out, shall either forcibly or privily relieve himself out of their abundance; especially if he do it with full intention to pay the value of them whenever his better fortune gives him ability. Some men deny that such a case of necessity, as we speak of, can possibly happen. But what if a man should wander in a foreign land, unknown, friendless, and in want, spoiled of all he had by shipwreck, or by robbers, or having lost by some casualty whatever he was worth in his own country; should none be found willing either to relieve his distress, or to hire his service, or should they rather (as it commonly happens,) seeing him in a good garb, suspect him to beg without reason, must the poor creature starve in this miserable condition?"

42. Many other great foreign authorities might be referred to, and I cannot help mentioning COVARRUVIUS, who is spoken of by JUDGE HALE, and who expresses himself upon the subject in these words: "The reason why a man in extreme necessity may, without incurring the guilt of theft or rapine, forcibly take the goods of others for his present relief, is because his condition renders all things common. For it is the ordinance and institution of nature itself, that inferior things should be designed and directed to serve the necessities of men. Wherefore the division of goods afterwards introduced into the world doth not derogate from that precept of natural reason, which Suggests, that the extreme wants of mankind may be in any manner removed by the use of temporal possessions." PUFFENDORF tells us, that PERESIUS maintains, that, in case of extreme necessity, a man is compelled to the action, by a force which he cannot resist; and then, that the owner's consent may be presumed on, because humanity obliges him to succour those who are in distress. The same writer cites a passage from St. AMBROSE, one of the FATHERS of the church, which alleges that (in case of refusing to give to persons in extreme necessity) it is the person who retains the goods who is guilty of the act of wrong doing, for St. AMBROSE says; "it is the bread of the hungry which you detain; it is the raiment of the naked which you lock up."

43. Before I come to the English authorities on the same side, let me again notice the foul dealing of Blackstone; let me point out another instance or two of the insincerity of this English court-sycophant, who was, let it be noted, Solicitor-general to the queen of the "good old King." You have seen, in paragraph 28, a most flagrant instance of his perversion of the Scriptures. He garbles the word of God, and prefaces the garbling by calling it a thing "certified by King Solomon himself;" and this word certified he makes use of just when he is about to begin the scandalous falsification of the text which he is referring to. Never was anything more base. But, the whole extent of the baseness we have not yet seen; for, BLACKSTONE had read HALE, who had quoted the two verses fairly; but besides this, he had read PUFFENDORF,

who had noticed very fully this text of Scripture, and who had shown very clearly that it did not at all make in favour of the doctrine of Blackstone. Blackstone ought to have given the argument of PUFFENDORF; he ought to have given the whole of his argument; but particularly he ought to have given this explanation of the passage in the PROVERBS, which explanation I have inserted in paragraph

27. It was also the height of insincerity in BLACKSTONE, to pretend that the passage from CICERO had anything at all to do with the matter. He knew well that it had not; he knew that CICERO contemplated no case of extreme necessity for want of food or clothing; but, he had read PUFFENDORF, and PUFFENDORF had told him, that CICERO'S was a question of the mere conveniences and inconveniences of life in general; and not a question of pinching hunger or shivering nakedness. BLACKSTONE had seen his fallacy exposed by PUFFENDORF; he had seen the misapplication of this passage of CICERO fully exposed by PUFFENDORF; and yet the base court-sycophant trumped it up again, without mentioning PUFFENDORF'S exposure of the fallacy! In short this BLACKSTONE, upon this occasion, as upon almost all others, has gone all lengths; has set detection and reproof at defiance, for the sake of making his court to the government by inculcating harshness in the application of the law, and by giving to the law such an interpretation as would naturally tend to justify that harshness.

44. Let us now cast away from us this insincere sycophant, and turn to other law authorities of our own country. The Mirrour of Justices, (quoted by me in paragraph 14,) Chap. 4, Section 16, on the subject of arrest of judgment of death, has this passage. Judgment is to be staid in seven cases here specified: and the seventh is this: "in POVERTY, in which case you are to distinguish of the poverty of the offender, or of things; for if poor people, to avoid famine, take victuals to sustain their lives, or clothes that they die not of cold, (so that they perish if they keep not themselves from cold,) they are not to be adjudged to death, if it were not in their power to have bought their victuals or clothes; for as much as they are warranted so to do by the law of nature." Now, my friends, you will observe, that I take this from a book which may almost be called the BIBLE of the law. There is no lawyer who will deny the goodness of this authority; or who will attempt to say that this was not always the law of England.

45. Our next authority is one quite as authentic, and almost as ancient. The book goes by the name of BRITTON, which was the name of a Bishop of Hereford, who edited it, in the famous reign of EDWARD THE FIRST. The book does, in fact, contain the laws of the kingdom as they existed at that time. It may be called the record of the laws of Edward the First. It begins thus, "Edward by the grace of God, King of England and Lord of Ireland, to all his liege subjects, peace, and grace of salvation." The preamble goes on to state, that people cannot be happy without good laws; that even good laws are of no use unless they be known and understood; and that, therefore, the king has ordered the laws of England thus to be written and recorded. This book is very well known to be of the greatest authority, amongst lawyers, and in Chap. 10 of this book, in which the law describes what constitutes a BURGLAR, or house-breaker, and the punishment that he shall suffer (which is that of death,) there is this passage: "Those are to be deemed burglars who feloniously, in time of peace, break into churches or houses, or through walls or doors of our cities, or our boroughs; with exception of children under age, and of poor people who for hunger, enter to take any sort of victuals of less value

than twelve pence; and except idiots and mad people, and others that cannot commit felony." Thus, you see, this agrees with the MIRROUR OF JUSTICES, and with all that we have read before from these numerous high authorities. But this, taken in its full latitude, goes a great length indeed; for a burglar is a breaker-in by night. So that this is not only a taking; but a breaking into a house in order to take! And observe, it is taking to the value of twelve pence; and twelve pence then was the price of a couple of sheep, and of fine fat sheep too; nay, twelve pence was the price of an ox, in this very reign of Edward the First. So that, a hungry man might have a pretty good belly-full in those days without running the risk of punishment. Observe, by-the-by, how time has hardened the law. We are told of the dark ages, of the barbarous customs, of our forefathers: and we have a SIR JAMES MACKINTOSH to receive and to present petitions innumerable, from the most tender hearted creatures in the world, about "softening the criminal code;" but, not a word do they ever say about a softening of this law, which now hangs a man for stealing the value of a RABBIT, and which formerly did not hang him till he stole the value of an OX! Curious enough, but still more scandalous, that we should have the impudence to talk of our humanity, and our civilization, and of the barbarousness of our forefathers. But, if a part of the ancient law remain, shall not the whole of it remain? If we hang the thief, still hang the thief for stealing to the value of twelve pence; though the twelve pence now represents a rabbit instead of an ox; if we still do this, would BLACKSTONE take away the benefit of the ancient law from the starving man? The passage that I have quoted is of such great importance as to this question, that I think it necessary to add, here, a copy of the original, which is in the old Norman-French, of which I give the translation above. "Sunt tenus burgessours trestous ceux, que felonisement en temps de pees debrusent esglises ou auter mesons, ou murs, ou portes de nos cytes, ou de nos burghes; hors pris enfauntz dedans age, et poures, que, pur feyn, entrêt pur ascun vitaille de meindre value q'de xii deners, et hors pris fous nastres, et gens arrages, et autres que seuent nule felonie faire."

46. After this, lawyers, at any rate, will not attempt to gainsay. If there should, however, remain any one to affect to doubt of the soundness of this doctrine, let them take the following from him who is always called the "pride of philosophy," the "pride of English learning," and whom the poet POPE calls "greatest and wisest of mankind." It is LORD BACON of whom I am speaking. He was Lord High Chancellor in the reign of James the First; and, let it be observed, that he wrote those "law tracts," from which I am about to quote, long after the present poor-laws had been established. He says (Law Tracts, page 55,) "The law chargeth no man with default where the act is compulsory and not voluntary, and where there is not consent and election; and, therefore, if either there be an impossibility for a man to do otherwise, or so great a perturbation of the judgment and reason, as in presumption of law a man's nature cannot overcome, such necessity carrieth a privilege in itself.--Necessity is of three sorts: necessity of conservation of life; necessity of obedience; and necessity of the act of God or of a stranger.--First, of conservation of life; if a man steal viands (victuals) to satisfy his present hunger, this is no felony nor larceny."

47. If any man want more authority, his heart must be hard indeed; he must have an uncommonly anxious desire to take away by the halter the life that sought to preserve itself against hunger. But, after all, what need had we of any authorities? What need had we even of reason upon the subject? Who is there upon the face of the earth, except the monsters that come from across the channel of St. George; who is there upon the face of the earth, except those

monsters, that have the brass, the hard hearts and the brazen faces, which enable them coolly to talk of the "MERIT" of the degraded creatures, who, amidst an abundance of food, amidst a "superabundance of food," lie quietly down and receive the extreme unction, and expire with hunger? Who, upon the face of the whole earth, except these monsters, these ruffians by way of excellence; who, except these, the most insolent and hard-hearted ruffians that ever lived, will contend, or will dare to think, that there ought to be any force under heaven to compel a man to lie down at the door of a baker's and butcher's shop, and expire with hunger! The very nature of man makes him shudder at the thought. There want no authorities; no appeal to law books; no arguments; no questions of right or wrong: that same human nature that tells me that I am not to cut my neighbour's throat, and drink his blood, tells me that I am not to make him die at my feet by keeping from him food or raiment of which I have more than I want for my own preservation.

48. Talk of barbarians, indeed; Talk of "the dark and barbarous ages." Why, even in the days of the DRUIDS, such barbarity as that of putting men to death, or of punishing them for taking to relieve their hunger, was never thought of. In the year 1811, the REV. PETER ROBERTS, A. M. published a book, entitled COLLECTANEA CAMBRICA. In the first volume of that book, there is an account of the laws of the ANCIENT BRITONS. Hume, and other Scotchmen, would make us believe, that the ancient inhabitants of this country were a set of savages, clothed in skins and the like. The laws of this people were collected and put into writing, in the year 694 before Christ. The following extract from these laws shows, that the moment civil society began to exist, that moment the law took care that people should not be starved to death. That moment it took care, that provision should be made for the destitute, or that, in cases of extreme necessity, men were to preserve themselves from death by taking from those who had to spare. The words of these laws (as applicable to our case) given by Mr. ROBERTS, are as follows:--"There are three distinct kinds of personal individual property, which cannot be shared with another, or surrendered in payment of fine; viz., a wife, a child, and argyfrew. By the word argyfrew is meant, clothes, arms, or the implements of a lawful calling. For without these a man has not the means of support, and it would be unjust in the law to unman a man, or to uncall a man as to his calling." TRIAD 53d.--"Three kinds of THIEVES are not to be punished with DEATH. 1. A wife, who joins with her husband in theft. 2. A youth under age. And 3. One who, after he has asked, in vain, for support, in three towns, and at nine houses in each town." TRIAD 137.

49. There were, then, houses and towns, it seems; and the towns were pretty thickly spread too; and, as to "civilization" and "refinement," let this law relative to a youth under age, be compared with the new orchard and garden law, and with the tread-mill affair, and new trespass law!

50. We have a law, called the VAGRANT ACT, to punish men for begging. We have a law to punish men for not working to keep their families. Now, with what show of justice can these laws be maintained? They are founded upon this; the first, that begging is disgraceful to the country; that it is degrading to the character of man, and, of course, to the character of an Englishman; and, that there is no necessity for begging, because the law has made ample provision for every person in distress. The law for punishing men for not working to maintain their families is founded on this, that they are doing wrong to their neighbours; their neighbours,

that is to say, the parish, being bound to keep the family, if they be not kept by the man's labour; and, therefore, his not labouring is a wrong done to the parish. The same may be said with regard to the punishment for not maintaining bastard children. There is some reason for these laws, as long as the poor-laws are duly executed; as long as the poor are duly relieved, according to law; but, unless the poor-laws exist; unless they be in full force; unless they be duly executed; unless efficient and prompt relief be given to necessitous persons, these acts, and many others approaching to a similar description, are acts of barefaced and most abominable tyranny. I should say that they would be acts of such tyranny; for generally speaking, the poor-laws are, as yet, fairly executed, and efficient as to their object.

51. The law of this country is, that every man, able to carry arms, is liable to be called on, to serve in the militia, or to serve as a soldier in some way or other, in order to defend the country. What, then, the man has no land; he has no property beyond his mere body, and clothes, and tools; he has nothing that an enemy can take away from him. What justice is there, then, in calling upon this man to take up arms and risk his life in the defence of the land: what is the land to him? I say, that it is something to him; I say, that he ought to be called forth to assist to defend the land; because, however poor he may be, he has a share in the land, through the poor-rates; and if he be liable to be called forth to defend the land, the land is always liable to be taxed for his support. This is what I say: my opinions are consistent with reason, with justice, and with the law of the land; but, how can MALTHUS and his silly and nasty disciples; how can those who want to abolish the poor-rates or to prevent the poor from marrying; how can this at once stupid and conceited tribe look the labouring man in the face, while they call upon him to take up arms, to risk his life, in defence of the land? Grant that the poor-laws are just; grant that every necessitous creature has a right to demand relief from some parish or other; grant that the law has most effectually provided that every man shall be protected against the effects of hunger and of cold; grant these, and then the law which compels the man without house or land to take up arms and risk his life in defence of the country, is a perfectly just law; but, deny to the necessitous that legal and certain relief of which I have been speaking; abolish the poor laws; and then this military-service law becomes an act of a character such as I defy any pen or tongue to describe.

52. To say another word upon the subject is certainly unnecessary; but we live in days when "stern necessity" has so often been pleaded for most flagrant departures from the law of the land, that one cannot help asking, whether there were any greater necessity to justify ADDINGTON for his deeds of 1817 than there would be to justify a starving man in taking a loaf? ADDINGTON pleaded necessity, and he got a Bill of Indemnity. And, shall a starving man be hanged, then, if he take a loaf to save himself from dying? When SIX ACTS were before the Parliament, the proposers and supporters of them never pretended that they did not embrace a most dreadful departure from the ancient laws of the land. In answer to LORD HOLLAND, who had dwelt forcibly on this departure from the ancient law, the Lord Chancellor, unable to contradict LORD HOLLAND, exclaimed, "Salus populi suprema lex," that is to say "The salvation of the people is the first law." Well, then, if the salvation of the people be the first law, the salvation of life is really and bona fide the salvation of the people; and, if the ordinary laws may be dispensed with, in order to obviate a possible and speculative danger, surely they may be dispensed with, in cases where to dispense with them is visibly, demonstrably, notoriously, necessary to the salvation of the lives of the people: surely, bread is as necessary to the lips of the starving man, as a new law could be necessary to prevent either house of parliament from

being brought into contempt; and surely, therefore, Salus populi suprema lex may come from the lips of the famishing people with as much propriety as they came from those of the Lord Chancellor!

53. Again, however, I observe, and with this I conclude, that we have nothing to do but to adhere to the poor-laws which we have; that the poor have nothing to do, but to apply to the overseer, or to appeal from him to the magistrate; that the magistrate has nothing to do but duly to enforce the law; and that the government has nothing to do, in order to secure the peace of the country, amidst all the difficulties that are approaching, great and numerous as they are; that it has nothing to do, but to enjoin on the magistrates to do their duty according to our excellent law; and, at the same time, the government ought to discourage, by all the means in their power, all projects for maintaining the poor by any other than legal means; to discourage all begging-box affairs; all miserable expedients; and also to discourage, and, where it is possible, fix its mark of reprobation upon all those detestable projectors, who are hatching schemes for what is called, in the blasphemous slang of the day, "checking the surplus population" who are hatching schemes for preventing the labouring people from having children: who are about spreading their nasty beastly publications; who are hatching schemes of emigration; and who, in short, seem to be doing every-thing in their power to widen the fearful breach that has already been made between the poor and the rich. The government has nothing to do but to cause the law to be honestly enforced; and then we shall see no starvation, and none of those dreadful conflicts which the fear of want, as well as actual want, never fail to produce. The bare thought of forced emigration to a foreign state, including, as it must, a transfer of all allegiance, which is contrary to the fundamental laws of England; or, exposing every emigrating person to the danger of committing high treason; the very thought of such a measure, having become necessary in England, is enough to make an Englishman mad. But, of these projects, these scandalous nasty beastly and shameless projects, we shall have time to speak hereafter; and in the mean while, I take my leave of you, for the present, by expressing my admiration of the sensible and spirited conduct of the people of STOCKPORT, when an attempt was, on the 5th of September, made to cheat them into an address, applauding the conduct of the Ministers! What! Had the people of STOCKPORT so soon forgotten 16th of August! Had they so soon forgotten their townsman, JOSEPH SWAN! If they had, they would have deserved to perish to all eternity. Oh, no! It was a proposition very premature: it will be quite soon enough for the good and sensible and spirited fellows of STOCKPORT; quite soon enough to address the Ministers, when the Ministers shall have proposed a repeal of the several Jubilee measures, called Ellenborough's law; the poacher-transporting law; the sun-set and sun-rise transportation law; the tread-mill law; the select-vestry law; the Sunday-toll laws; the new trespass law; the new treason law; the seducing-soldier-hanging law; the new apple-felony law; the SIX ACTS; and a great number of others, passed in the reign of Jubilee. Quite soon enough to applaud, that is, for the sensible people of STOCKPORT to applaud, the Ministers, when those Ministers have proposed to repeal these laws, and, also, to repeal the malt tax, and those other taxes, which take, even from the pauper, one half of what the parish gives him to keep the breath warm in his body. Quite soon enough to applaud the Ministers, when they have done these things; and when in addition to all these, they shall have openly proposed a radical reform of the Commons House of Parliament. Leaving them to do this as soon as they like, and trusting, that you will never, on any account, applaud them until they do it, I, expressing here my best thanks to Mr. BLACKSHAW, who defeated the

slavish scheme at Stockport, remain, Your faithful friend, and most obedient servant, WM. COBBETT.

NUMBER III.

Hurstbourne Tarrant (called Uphusband,)

Hants, 13th October, 1826.

MY EXCELLENT FRIENDS,

54. In the foregoing Numbers, I have shown, that men can never be so poor as to have no rights at all: and that, in England, they have a legal, as well as a natural, right to be maintained, if they be destitute of other means, out of the lands, or other property, of the rich. But, it is an interesting question, HOW THERE CAME TO BE SO MUCH POVERTY AND MISERY IN ENGLAND. This is a very interesting question; for, though it is the doom of man, that he shall never be certain of any-thing, and that he shall never be beyond the reach of calamity; though there always has been, and always will be, poor people in every nation; though this circumstance of poverty is inseparable from the means which uphold communities of men; though, without poverty, there could be no charity, and none of those feelings, those offices, those acts, and those relationships, which are connected with charity, and which form a considerable portion of the cement of civil society: yet, notwithstanding these things, there are bounds beyond which the poverty of the people cannot go, without becoming a thing to complain of, and to trace to the Government as a fault. Those bounds have been passed, in England, long and long ago. England was always famed for many things; but especially for its good living; that is to say, for the plenty in which the whole of the people lived; for the abundance of good clothing and good food which they had. It was always, ever since it bore the name of England, the richest and most powerful and most admired country in Europe; but, its good living, its superiority in this particular respect, was proverbial amongst all who knew, or who had heard talk of, the English nation. Good God! How changed! Now, the very worst fed and worst clad people upon the face of the earth, those of Ireland only excepted. How, then, did this horrible, this disgraceful, this cruel poverty come upon this once happy nation? This, my good friends of Preston, is, to us all, a most important question; and, now let us endeavour to obtain a full and complete answer to it.

55. POVERTY is, after all, the great badge, the never-failing badge, of slavery. Bare bones and rags are the true marks of the real slave. What is the object of Government? To cause men to live happily. They cannot be happy without a sufficiency of food and of raiment. Good government means a state of things in which the main body are well fed and well clothed. It is the chief business of a government to take care, that one part of the people do not cause the other part to lead miserable lives. There can be no morality, no virtue, no sincerity, no honesty, amongst a people continually suffering from want; and, it is cruel, in the last degree, to punish such people for almost any sort of crime, which is, in fact, not crime of the heart, not crime of the perpetrator, but the crime of his all-controlling necessities.--To what degree the main body of the people, in England, are now poor and miserable; how deplorably wretched they now are; this we know but too well; and now, we will see what was their state before this vaunted "REFORMATION." I shall be very particular to cite my authorities here. I will infer nothing; I

will give no "estimate;" but refer to authorities, such as no man can call in question, such as no man can deny to be proofs more complete than if founded on oaths of credible witnesses, taken before a judge and jury. I shall begin with the account which FORTESCUE gives of the state and manner of living of the English, in the reign of Henry VI.; that is, in the 15th century, when the Catholic Church was in the height of its glory. FORTESCUE was Lord Chief Justice of England for nearly twenty years; he was appointed Lord High Chancellor by Henry VI. Being in exile, in France, in consequence of the wars between the Houses of York and Lancaster, and the King's son, Prince Edward, being also in exile with him, the Chancellor wrote a series of Letters, addressed to the Prince, to explain to him the nature and effects of the Laws of England, and to induce him to study them and uphold them. This work, which was written in Latin, is called De Laudibus Legum Angliæ; or, PRAISE OF THE LAWS OF ENGLAND. This book was, many years ago, translated into English, and it is a book of Law-Authority, quoted frequently in our courts of this day. No man can doubt the truth of facts related in such a work. It was a work written by a famous lawyer for a prince; it was intended to be read by other contemporary lawyers, and also by all lawyers in future. The passage that I am about to quote, relating to the state of the English, was purely incidental; it was not intended to answer any temporary purpose. It must have been a true account.--The Chancellor, after speaking generally of the nature of the laws of England, and of the difference between them and the laws of France, proceeds to show the difference in their effects, by a description of the state of the French people, and then by a description of the state of the English. His words, words that, as I transcribe them, make my cheeks burn with shame, are as follows: "Besides all this, the inhabitants of France give every year to their King the fourth part of all their wines, the growth of that year, every vintner gives the fourth penny of what he makes of his wine by sale. And all the towns and boroughs pay to the King yearly great sums of money, which are assessed upon them, for the expenses of his men at arms. So that the King's troops, which are always considerable, are substituted and paid yearly by those common people, who live in the villages, boroughs, and cities. Another grievance is, every village constantly finds and maintains two cross-bow-men, at the least; some find more, well arrayed in all their accoutrements, to serve the King in his wars, as often as he pleaseth to call them out, which is frequently done. Without any consideration had of these things, other very heavy taxes are assessed yearly upon every village within the kingdom, for the King's service; neither is there ever any intermission or abatement of taxes. Exposed to these and other calamities, the peasants live in great hardship and misery. Their constant drink is water, neither do they taste, throughout the year, any other liquor, unless upon some extraordinary times, or festival days. Their clothing consists of frocks, or little short jerkins, made of canvass, no better than common sackcloth; they do not wear any woollens, except of the coarsest sort; and that only in the garment under their frocks; nor do they wear any trowse, but from the knees upwards; their legs being exposed and naked. The women go barefoot, except on holidays. They do not eat flesh, except it be the fat of bacon, and that in very small quantities, with which they make a soup. Of other sorts, either boiled or roasted, they do not so much as taste, unless it be of the inwards and offals of sheep and bullocks, and the like which are killed, for the use of the better sort of people, and the merchants; for whom also quails, partridges, hares, and the like, are reserved, upon pain of the gallies; as for their poultry, the soldiers consume them, so that scarce the eggs, slight as they are, are indulged them, by way of a dainty. And if it happen that a man is observed to thrive in the world, and become rich, he is presently assessed to the King's tax, proportionably more than his poorer neighbours, whereby he is soon reduced to a level with the rest." Then comes his description of the ENGLISH, at the same time; those "priest-ridden"

English, whom CHALMERS and HUME, and the rest of that tribe, would fain have us believe, were a mere band of wretched beggars.--"The King of England cannot alter the laws, or make new ones, without the express consent of the whole kingdom in Parliament assembled. Every inhabitant is at his liberty fully to use and enjoy whatever his farm produceth, the fruits of the earth, the increase of his flock, and the like: all the improvements he makes, whether by his own proper industry, or of those he retains in his service, are his own, to use and enjoy, without the let, interruption, or denial of any. If he be in anywise injured or oppressed, he shall have his amends and satisfactions against the party offending. Hence it is that the inhabitants are rich in gold, silver, and in all the necessaries and conveniences of life. They drink no water, unless at certain times, upon a religious score, and by way of doing penance. They are fed, in great abundance, with all sorts of flesh and fish, of which they have plenty every-where; they are clothed throughout in good woollens; their bedding and other furniture in their houses are of wool, and that in great store. They are also well provided with all other sorts of household goods and necessary implements for husbandry. Every one, according to his rank, hath all things which conduce to make life easy and happy."--Go, and read this to the poor souls, who are now eating sea-weed in Ireland; who are detected in robbing the pig-troughs in Yorkshire; who are eating horse-flesh and grains (draff) in Lancashire and Cheshire; who are harnessed like horses, and drawing gravel in Hampshire and Sussex; who have 3d. a day allowed them by the magistrates in Norfolk; who are, all over England, worse fed than the felons in the jails. Go, and tell them, when they raise their hands from the pig-trough, or from the grains-tub, and, with their dirty tongues, cry "No Popery;" go, read to the degraded and deluded wretches, this account of the state of their Catholic forefathers, who lived under what is impudently called "Popish superstition and tyranny," and in those times which we have the audacity to call "the dark ages."--Look at the then picture of the French; and, Protestant Englishmen, if you have the capacity of blushing left, blush at the thought of how precisely that picture fits the English now! Look at all the parts of the picture; the food, the raiment, the game! Good God! If any one had told the old Chancellor, that the day would come, when this picture, and even a picture more degrading to human nature, would fit his own boasted country, what would he have said? What would he have said, if he had been told, that the time was to come, when the soldier, in England, would have more than twice, nay, more than thrice, the sum allowed to the day-labouring man; when potatoes would be carried to the field as the only food of the ploughman; when soup-shops would be open to feed the English; and when the Judges, sitting on that very Bench on which he himself had sitten for twenty years, would (as in the case of last year of the complaints against Magistrates at NORTHALLERTON) declare that BREAD AND WATER were the general food of working people in England? What would he have said?

Why, if he had been told, that there was to be a "REFORMATION," accompanied by a total devastation of Church and Poor property, upheld by wars, creating an enormous Debt and enormous taxes, and requiring a constantly standing army; if he had been told this, he would have foreseen our present state, and would have wept for his country; but, if he had, in addition, been told, that, even in the midst of all this suffering, we should still have the ingratitude and the baseness to cry "No Popery," and the injustice and the cruelty to persecute those Englishmen and Irishmen, who adhered to the faith of their pious, moral, brave, free and happy fathers, he would have said, "God's will be done: let them suffer."--But, it may be said, that it was not, then, the Catholic Church, but the Laws, that made the English so happy; for, the French had that Church as well as the English.

Aye! But, in England, the Church was the very basis of the laws. The very first clause of MAGNA CHARTA provided for the stability of its property and rights. A provision for the indigent, an effectual provision, was made by the laws that related to the Church and its property; and this was not the case in France; and never was the case in any country but this: so that the English people lost more by a "Reformation" than any other people could have lost.--Fortescue's authority would, of itself, be enough; but, I am not to stop with it. WHITE, the late Rector of SELBOURNE, in Hampshire, gives, in his History of that once-famous village, an extract from a record, stating that for disorderly conduct, men were punished by being "compelled to fast a fortnight on bread and beer!" This was about the year 1380, in the reign of RICHARD II. Oh! miserable "dark ages!" This fact must be true.

WHITE had no purpose to answer. His mention of the fact, or rather his transcript from the record, is purely incidental; and trifling as the fact is, it is conclusive as to the general mode of living in those happy days. Go, tell the harnessed gravel-drawers, in Hampshire, to cry "No Popery;" for, that, if the Pope be not put down, he may, in time, compel them to fast on bread and beer, instead of suffering them to continue to regale themselves on nice potatoes and pure water.--But, let us come to Acts of Parliament, and, first, to the Act above mentioned of KING EDWARD III. That Act fixes the price of meat. After naming the four sorts of meat, beef, pork, mutton, and veal, the preamble has these words: "These being THE FOOD OF THE POORER SORT." This is conclusive. It is an incidental mention of a fact.

It is an Act of Parliament. It must have been true; and, it is a fact that we know well, that even the Judges have declared from the Bench, that bread alone is now the food of the poorer sort. What do we want more than this to convince us, that the main body of the people have been impoverished by the "Reformation?"--But I will prove, by other Acts of Parliament, this Act of Parliament to have spoken truth. These Acts declare what the wages of workmen shall be. There are several such Acts, but one or two may suffice. The Act of 23d of EDW. III. fixes the wages, without food, as follows.

There are many other things mentioned, but the following will be enough for our purpose. s. d.

A woman hay-making, or weeding corn, for the day 0 1 A man filling dung-cart 0 3-1/2 A reaper 0 4 Mowing an acre of grass 0 6 Thrashing a quarter of Wheat 0 4 The price of shoes, cloth, and of provisions, throughout the time that this law continued in force, was as follows:--

L. s. d.

A pair of shoes 0 0 4 Russet broad-cloth the yard 0 1 1 A stall-fed ox 1 4 0 A grass-fed ox 0 16 0 A fat sheep unshorn 0 1 8 A fat sheep shorn 0 1 2 A fat hog 2 years old 0 3 4 A fat goose 0 0 2-1/2 Ale, the gallon, by proclamation 0 0 1 Wheat the quarter 0 3 4 White wine the gallon 0 0 6 Red wine 0 0 4 These prices are taken from the PRECIOSUM of BISHOP FLEETWOOD, who took them from the accounts kept by the bursers of convents. All the world knows, that FLEETWOOD'S book is of undoubted authority.--We may then easily believe, that "beef, pork, mutton, and veal," were "the food of the poorer sort," when a dung-cart filler had more than the price of a fat goose and a half for a day's work, and when a woman was allowed, for a day's weeding, the price of a quart of red wine! Two yards of the cloth made a coat for the shepherd; and, as it cost 2s. 2d., the reaper would earn it in 6-1/2 days; and, the dung-cart man would earn very nearly a pair of shoes every day! this dung-cart filler would earn a fat shorn sheep in four days; he would earn a fat hog, two years old, in twelve days; he would earn a grass-fed ox in twenty days; so that we may easily believe, that "beef, pork, and mutton," were "the food of the poorer sort." And, mind, this was "a priest-ridden people;" a people "buried in Popish superstition!" In our days of "Protestant light" and of "mental enjoyment," the "poorer sort" are allowed by the Magistrates of Norfolk, 3d. a day for a single man able to work. That is to say, a half-penny less than the Catholic dung-cart man had; and that 3d. will get the "No Popery" gentleman about six ounces of old ewe-mutton, while the Popish dung-cart man got, for his day, rather more than the quarter of a fat sheep.--But, the popish people might work harder than "enlightened Protestants." They might do more work in a day. This is contrary to all the assertions of the feelosophers; for they insist, that the Catholic religion made people idle. But, to set this matter at rest, let us look at the price of the job-labour; at the mowing by the acre, and at the thrashing of wheat by the quarter; and let us see how these wages are now, compared with the price of food. I have no parliamentary authority since the year 1821, when a report was printed by order of the House of Commons, containing the evidence of Mr. ELLMAN, of Sussex, as to wages, and of Mr. GEORGE, of Norfolk, as to price of wheat. The report was dated 18th June, 1821. The accounts are for 20 years, on an average, from 1800 inclusive. We will now proceed to see how the "popish, priest-ridden" Englishman stands in comparison with the "No Popery" Englishman.

POPISH MAN. NO POPERY MAN.

s. d. s. d.

Mowing an acre of grass 0 6 3 7-3/4 Thrashing a quarter of Wheat 0 4 4 0 Here are "waust improvements, Mau'm!" But, now let us look at the relative price of the wheat, which the labourer had to purchase with his wages. We have seen, that the "popish superstition slave" had to give fivepence a bushel for his wheat, and the evidence of Mr. GEORGE states, that the "enlightened Protestant" had to give 10 shillings a bushel for his wheat; that is 24 times as much as the "popish fool," who suffered himself to be "priest-ridden." So that the "enlightened" man, in order to make him as well off as the "dark-ages" man was, ought to receive twelve shillings, instead of 3s. 7-3/4d. for mowing an acre of grass; and he, in like manner, ought to receive, for thrashing a quarter of wheat, eight shillings, instead of the four shillings which he does receive. If we had the records, we should doubtless find, that IRELAND was in the same state.

56. There! That settles the matter as to ancient good living. Now, as to the progress of poverty and misery, amongst the working people, during the last half century, take these facts; in the year 1771, that is, 55 years ago, ARTHUR YOUNG, who was afterwards Secretary to the Board of Agriculture, published a work on the state of the agriculture of the country, in which he gave the allowance for the keeping of a farm-labourer, his wife and three children, which allowance, reckoning according to the present money-price of the articles which he allows amounted to 13s. 1d. He put the sum, at what he deemed the lowest possible sum, on which the people could exist. Alas! we shall find, that they can be made to exist upon little more than one-half of this sum!

57. This allowance of Mr. ARTHUR YOUNG was made, observe, in 1771, which was before the Old American War took place. That war made some famous fortunes for admirals and commodores and contractors and pursers and generals and commissaries; but, it was not the Americans, the French, nor the Dutch, that gave the money to make these fortunes. They came out of English taxes; and the heaviest part of those taxes fell upon the working people, who, when they were boasting of "victories," and rejoicing that the "JACK TARS" had got "prize-money," little dreamed that these victories were purchased by them, and that they paid fifty pounds for every crown that sailors got in prize-money! In short, this American war caused a great mass of new taxes to be laid on, and the people of England became a great deal poorer than they ever had been before. During that war, they BEGAN TO EAT POTATOES, as something to "save bread." The poorest of the people, the very poorest of them, refused, for a long while, to use them in this way; and even when I was ten years old, which was just about fifty years ago; the poor people would not eat potatoes, except with meat, as they would cabbages, or carrots, or any other moist vegetable. But, by the end of the American war, their stomachs had come to! By slow degrees they had been reduced to swallow this pig-meat, (and bad pig-meat too,) not, indeed, without grumbling; but to swallow it; to be reduced, thus, many degrees in the scale of animals.

58. At the end of twenty-four years from the date of ARTHUR YOUNG'S allowance, the poverty and degradation of the English people had made great strides. We were now in the year 1795, and a new war, and a new series of "victories and prizes" had begun. But who it was that suffered for these, out of whose blood and flesh and bones they came, the allowance now (in 1795) made to the poor labourers and their families will tell. There was, in that year, a TABLE, or SCALE, of allowance, framed by the Magistrates of Berkshire. This is, by no means, a hard county; and therefore it is reasonable to suppose, that the scale was as good a one for the poor as any in England. According to this scale, which was printed and published, and also acted upon for years, the weekly allowance, for a man, his wife and three children, was, according to present money-prices, 11s. 4d. Thus it had, in the space of twenty-four years, fell from 13s. 1d. to 11s. 4d. Thus were the people brought to the pig-meat! Food, fit for men, they could not have with 11s. 4d. a week for five persons.

59. One would have thought, that to make a human being live upon 4d. a day, and find fuel, clothing, rent, washing, and bedding, out of the 4d., besides eating and drinking, was impossible; and one would have thought it impossible for any-thing not of hellish birth and breeding, to entertain a wish to make poor creatures, and our neighbours too, exist in such a state of horrible misery and degradation as the labourers of England were condemned to by this scale of 1795. Alas! this was happiness and honour; this was famous living; this 11s. 4d. a week was luxury and feasting, compared to what we NOW BEHOLD! For now the allowance, according to present money-prices, is 8s. a week for the man, his wife, and three children; that is to say 2-5/7 d. In words, TWO PENCE AND FIVE SEVENTHS OF ANOTHER PENNY, FOR A DAY! There, that is England now! That is what the base wretches, who are fattening upon the people's labour, call "the envy of surrounding nations and the admiration of the world." This is what SIR FRANCIS BURDETT applauds; and he applauds the mean and cruel and dastardly ruffians, whom he calls, "the country gentlemen of England," and whose generosity he cries up; while he well knows, that it is they (and he amongst the rest) who are the real and only cause of this devil-like barbarity, which (and he well knows that too) could not possibly be

practised without the constant existence and occasional employment of that species of force, which is so abhorrent to the laws of England, and of which this Burdett's son forms a part. The poor creatures, if they complain; if their hunger make them cry out, are either punished by even harder measures, or are slapped into prison. Alas! the jail is really become a place of relief, a scene of comparative good living: hence the invention of the tread-mill! What shall we see next? Workhouses, badges, hundred-houses, select-vestries, tread-mills, gravel-carts, and harness! What shall we see next! And what should we see at last, if this infernal THING could continue for only a few years longer?

60. In order to form a judgment of the cruelty of making our working neighbours live upon 2-5/7d. a day; that is to say 2d. and rather more than a halfpenny, let us see what the surgeons allow in the hospitals, to patients with broken limbs, who, of course, have no work to do, and who cannot even take any exercise. In GUY'S HOSPITAL, London, the daily allowance to patients, having simple fractures, is this: 6 ounces of meat; 12 ounces of bread; 1 pint of broth; 2 quarts of good beer. This is the daily allowance. Then, in addition to this, the same patient has 12 ounces of butter a week. These articles, for a week, amount to not less at present retail prices (and those are the poor man's prices,) than 6s. 9d. a week; while the working man is allowed 1s. 7d. a week! For, he cannot and he will not see his wife and children actually drop down dead with hunger before his face; and this is what he must see, if he take to himself more than a fifth of the allowance for the family.

61. Now, pray, observe, that surgeons, and particularly those eminent surgeons who frame rules and regulations for great establishments like that of Guy's Hospital, are competent judges of what nature requires in the way of food and of drink. They are, indeed, not only competent judges, but they are the best of judges: they know precisely what is necessary; and having the power to order the proper allowance, they order it. If, then, they make an allowance like that, which we have seen, to a person who is under a regimen for a broken limb; to a person who does no work, and who is, nine times out of ten, unable to take any exercise at all, even that of walking about, at least in the open air; if the eminent surgeons of London deem six shillings and ninepence worth of victuals and drink, a week, necessary to such a patient; if they think that nature calls for so much in such a case; what must that man be made of, who can allow to a working man, a man fourteen hours every day in the open air, one shilling and seven pence worth of victuals and drink for the week! Let me not however ask what "that man" can be made of; for it is a monster and not a man: it is a murderer of men: not a murderer with the knife or the pistol, but with the more cruel instrument of starvation. And yet, such monsters go to church and to meeting; aye, and subscribe, the base hypocrites, to circulate that Bible which commands to do as they would be done by, and which, from the first chapter to the last, menaces them with punishment, if they be hard to the poor, the fatherless, the widow, or the stranger!

62. But, not only is the patient, in a hospital, thus so much more amply fed than the working man; the prisoners in the jails; aye, even the convicted felons, are fed better, and much better, than the working men now are! Here is a fine "Old England;" that country of "roast beef and plumb pudding:" that, as the tax-eaters say it is, "envy of surrounding nations and admiration of the world." Aye; the country WAS all these; but, it is now precisely the reverse of them all. We have just seen that the honest labouring man is allowed 2-5/7d. a day; and that will buy him a pound and a half of good bread a day, and no more, not a single crumb more. This is all he

has. Well enough might the Hampshire Baronet, SIR JOHN POLLEN, lately, at a meeting at Andover, call the labourers "poor devils," and say, that they had "scarcely a rag to cover them!" A pound and a half of bread a day, and nothing more, and that, too, to work upon! Now, then, how fare the prisoners in the jails? Why, if they be CONVICTED FELONS, they are, say the Berkshire jail-regulations, "to have ONLY BREAD and water, with vegetables occasionally from the garden." Here, then, they are already better fed than the honest labouring man. Aye, and this is not all; for, this is only the week-day fare; for, they are to have, "on Sundays, SOME MEAT and broth!" Good God! And the honest working man can never, never smell the smell of meat! This is "envy of surrounding nations" with the devil to it! This is a state of things for Burdett to applaud.

63. But we are not even yet come to a sight of the depth of our degradation. These Berkshire jail-regulations make provision for setting the convicted prisoners, in certain cases, TO WORK, and, they say, "if the surgeon think it necessary, the WORKING PRISONERS may be allowed MEAT AND BROTH ON WEEK-DAYS;" and on Sundays, of course! There it is! There is the "envy and admiration!" There is the state to which Mr. Prosperity and Mr. Canning's best Parliament has brought us. There is the result of "victories" and prize-money and battles of Waterloo and of English ladies kissing, "Old Blucher." There is the fruit, the natural fruit, of anti-jacobinism and battles on the Serpentine River and jubilees and heaven-born ministers and sinking-funds and "public credit" and army and navy contracts. There is the fruit, the natural, the nearly (but not quite) ripe fruit of it all: the CONVICTED FELON is, if he do not work at all, allowed, on week-days, some vegetables in addition to his bread, and, on Sundays, both meat and broth; and, if the CONVICTED FELON work, if he be a WORKING convicted felon, he is allowed meat and broth all the week round; while, hear it Burdett, thou Berkshire magistrate! hear it, all ye base miscreants who have persecuted men because they sought a reform!

The WORKING CONVICTED FELON is allowed meat and broth every day in the year, while the WORKING HONEST MAN is allowed nothing but dry bread, and of that not half a belly-full! And yet you see the people that seem surprised that crimes increase! Very strange, to be sure; that men should like to work upon meat and broth better than they like to work upon dry bread! No wonder that new jails arise. No wonder that there are now two or three or four or five jails to one county, and that as much is now written upon "prison discipline" as upon almost any subject that is going.

But, why so good, so generous, to FELONS? The truth is, that they are not fed too well; for, to be starved is no part of their sentence; and, here are SURGEONS who have something to say! They know very well that a man may be murdered by keeping necessary food from him. Felons are not apt to lie down and die quietly for want of food. The jails are in large towns, where the news of any cruelty soon gets about. So that the felons have many circumstances in their favour. It is in the villages, the recluse villages, where the greatest cruelties are committed.

64. Here, then, in this contrast between the treatment of the WORKING FELON and that of the WORKING HONEST MAN, we have a complete picture of the present state of England; that horrible state, to which, by slow degrees, this once happy country has been brought; and, I should now proceed to show, as I proposed in the first paragraph of this present Number, HOW THERE CAME TO BE SO MUCH POVERTY AND MISERY IN ENGLAND; for, this is the main thing, it being clear, that, if we do not see the real causes of

our misery, we shall be very unlikely to adopt any effectual remedy. But, before I enter on this part of my subject, let me prove, beyond all possibility of doubt, that what I say relatively to the situation of, and the allowances to, the labourers and their families, IS TRUE. The cause of such situation and allowances I shall show hereafter; but let me first show, by a reference to indubitable facts, that the situation and allowances are such as, or worse than, I have described them. To do this, no way seems to me to be so fair, so likely to be free from error, so likely to produce a suitable impression on the minds of my readers, and so likely to lead to some useful practical result; no way seems to me so well calculated to answer these purposes, as that of taking the very village, in which, I, at this moment, happen to be, and to describe, with names and dates, the actual state of its labouring people, as far as that state is connected with steps taken under the poor-laws.

65. This village was in former times a very considerable place, as is manifest from the size of the church as well as from various other circumstances. It is now, as a church living, united with an adjoining parish, called VERNON DEAN, which also has its church, at a distance of about three miles from the church of this parish. Both parishes put together now contain only eleven hundred, and a few odd, inhabitants, men, women, children, and all; and yet, the great tithes are supposed to be worth two or three thousand pounds a year, and the small tithes about six hundred pounds a year. Formerly, before the event which is called "THE REFORMATION," there were two Roman Catholic priests living at the parsonage houses in these two parishes. They could not marry, and could, therefore have no wives and families to keep out of the tithes; and, WITH PART OF THOSE TITHES, THEY, AS THE LAW PROVIDED, MAINTAINED THE POOR OF THESE TWO PARISHES; and, the canons of the church commanded them to distribute the portion to the poor and the stranger, "with their own hands, in humility and mercy."

66. This, as to church and poor, was the state of these villages, in the "dark ages" of "Romish superstition." What! No poor-laws? No poor-rates? What horribly unenlightened times! No select vestries? Dark ages indeed! But, how stands these matters now? Why, the two parishes are moulded into one church living. Then the GREAT TITHES (amounting to two or three thousand a year) belong to some part of the Chapter (as they call it) of Salisbury. The Chapter leases them out, as they would a house or a farm, and they are now rented by JOHN KING, who is one of this happy nation's greatest and oldest pensioners. So that, away go the great tithes, not leaving a single wheat-ear to be spent in the parish. The SMALL TITHES belong to a VICAR, who is one FISHER, a nephew of the late bishop of Salisbury, who has not resided here for a long while; and who has a curate, named JOHN GALE, who being the son of a little farmer and shop-keeper at BURBAGE in Wiltshire, was, by a parson of the name of BAILEY (very well known and remembered in these parts), put to school; and, in the fulness of time, became a curate. So that, away go also the small tithes (amounting to about 500l. or 600l. a year); and, out of the large church revenues; or, rather, large church-and-poor revenues, of these two parishes; out of the whole of them, there remains only the amount of the curate, Mr. JOHN GALE'S, salary, which does not, perhaps, exceed seventy or a hundred pounds, and a part of which, at any rate, I dare say, he does not expend in these parishes: away goes, I say, all the rest of the small tithes, leaving not so much as a mess of milk or a dozen of eggs, much less a tithe-pig, to be consumed in the parish.

67. As to the poor, the parishes continue to be in two; so that I am to be considered as speaking of the parish of UPHUSBAND only. You are aware, that, amongst the last of the acts of the famous JUBILEE-REIGN, was an act to enable parishes to establish SELECT VESTRIES; and one of these vestries now exists in this parish. And now, let me explain to you the nature and tendency of this Jubilee-Act. Before this Act was passed, overseers of the poor had full authority to grant relief at their discretion. Pray mark that. Then again, before this Act was passed, any one justice of the peace might, on complaint of any poor person, order relief. Mark that. A select vestry is to consist of the most considerable rate-payers. Mark that. Then, mark these things: this Jubilee-Act forbids the overseer to grant any relief other than such as shall be ordered by the select vestry: it forbids ONE justice to order relief, in any case, except in a case of emergency: it forbids MORE THAN ONE to order relief, except on oath that the complainant has applied to the select vestry (where there is one,) and has been refused relief by it; and that, in no case, the justice's order shall be for more than a month; and, moreover, that when a poor person shall appeal to justices from a select vestry, the justices, in ordering relief, or refusing, shall have "regard to the conduct and CHARACTER of the applicant!"

68. From this Act, one would imagine, that overseers and justices were looked upon as being too soft and yielding a nature; too good, too charitable, too liberal to the poor! In order that the select vestry may have an agent suited to the purposes that the Act manifestly has in view, the Act authorizes the select vestry to appoint what is called an "assistant overseer," and to give him a salary out of the poor-rates. Such is this Jubilee-Act, one of the last Acts of the Jubilee-reign, that reign, which gave birth to the American war, to Pitt, to Perceval, Ellenborough, Sidmouth, and Castlereagh, to a thousand millions of taxes and another thousand millions of debt: such is the Select Vestry Act; and this now little trifling village of UPHUSBAND has a Select-Vestry! Aye, and an "ASSISTANT OVERSEER," too, with a salary of FIFTY POUNDS A YEAR, being, as you will presently see, about a SEVENTH PART OF THE WHOLE OF THE EXPENDITURE ON THE POOR!

69. The Overseers make out and cause to be printed and published, at the end of every four weeks, an account of the disbursements. I have one of these accounts now before me; and I insert it here, word for word, as follows:--

70. "The disbursements of Mr. T. Child and Mr. C. Church, bread at 1s. 2d. per gallon. Sept. 25th, 1826.

WIDOWS.

£. s. d. £. s. d. Blake, Ann 0 8 0 Bray, Mary 0 8 0 Cook, Ann 0 7 6 Clark, Mary 0 10 0 Gilbert, Hannah 0 8 0 Marshall, Sarah 0 10 0 Smith, Mary 0 8 0 Westrip, Jane 0 8 0 Withers, Ann 0 8 0 Dance, Susan 0 8 0 --------- 4 3 6 BASTARDS. ---- ---- 0 7 0 ---- ---- 0 6 0 ---- ---- 0 7 0 --- - ---- 0 6 0 ---- ---- 2 children 0 12 0 ---- ---- 2 children 0 12 0 ---- ---- - 10 0 ---- ---- - 8 0 ---- ---- - 6 0 ---- ---- - 8 0 ---- ---- - 8 0 ---- ---- - 6 0 ---- ---- - 6 0 ---- ---- - 6 0 --------- 5 8 0

OLD MEN.

Blake, John 0 16 0 Cannon, John 0 14 0 Cummins, Peter 0 16 0 Hopgood, John 0 16 0 Holden, William 0 6 0 Marshall, Charles 0 16 0 Nutley, George 0 7 0 --------- 4 11 0

FAMILIES.

Bowley, Mary 0 4 0 Baverstock, Elizabeth, 2 children 0 9 4 Cook, Levi 5 children 0 5 4 Kingston, John 6 ditto 0 10 0 Knight, John 6 ditto 0 10 0 Newman, David 5 ditto 0 5 4 Pain, Robert 5 ditto 0 5 4 Synea, William 6 ditto 0 10 0 Smith, Sarah (Moses) 1 ditto 0 4 8 Studman, Sarah 2 ditto 0 9 4 White, Joseph 8 ditto 0 19 4 Wise, William 6 ditto 0 10 0 Waldren, Job 5 ditto 0 5 4 Noyce, M. Batt, 7do. 6 weeks' pay 1 2 0 --------- 6 10 0 EXTRA IN THIS MONTH.

Thomas Farmer, ill 3 days 0 4 0 Levi Cook, ill 4 weeks and 1 day 1 13 4 Joseph White's child, 6 weeks 0 7 0 Jane Westrip's rent 0 2 0 William Fisher, 1 month ill 1 12 0 Paid boy, 2 days ill 0 0 8 James Orchard, ill 1 0 2 James Orchard's daughter, ill 0 8 0 Adders and Sparrows 0 2 3-1/2 Wicks for Carriage 0 1 0 Paid Mary Hinton 0 4 0 Joseph Farmer, ill 3 days 0 2 9 Thomas Cummins 0 6 0 Samuel Day, and son, ill 0 8 2 --------- 6 11 4 Total amount for the 4 weeks 27 3 10-1/2

71. Under the head of "WIDOWS" are, generally, old women wholly unable to work; and that of "OLD MEN" are men past all labour: in some of the instances lodging places, in very poor and wretched houses, are found these old people, and, in other instances, they have the bare money; and, observe, that money is FOR FOUR WEEKS! Gracious God! Have we had no mothers ourselves! Were we not born of woman! Shall we not feel then for the poor widow who, in her old age, is doomed to exist on two shillings a week, or threepence halfpenny a day, and to find herself clothes and washing and fuel and bedding out of that! And, the poor old men, the very happiest of whom gets, you see, less than 7d. a day, at the end of 70 or 80 years of a life, all but six of which have been years of labour! I have thought it right to put blanks instead of the names, under the second head.

Men of less rigid morality, and less free from all illicit intercourse, than the members of the Select Vestry of Uphusband, would, instead of the word "bastard," have used the more amiable one of "love-child;" and, it may not be wholly improper to ask these rigid moralists, whether they be aware, that they are guilty of LIBEL, aye, of real criminal libel, in causing these poor girls' names to be printed and published in this way. Let them remember, that the greater the truth the greater the libel; and, let them remember, that the mothers and the children too, may have memories! But, it is under the head of "FAMILIES" that we see that which is most worthy of our attention. Observe, that eight shillings a week is the wages for a day labourer in the village. And, you see, it is only when there are more than four children that the family is allowed anything at all. "LEVI COOK," for instance, has five children, and he receives allowance for one child. "JOSEPH WHITE" has eight children, and he receives allowance for four. There are three widows under this head; but, it is where there is a man, the father of the family, that we ought to look with attention; and here we find, that nothing at all is allowed to a family of a man, a wife, and four children, beyond the bare eight shillings a week of wages; and this is even worse than the allowance which I contrasted with that of the hospital patients and convicted felons; for there I supposed the family to consist of a man, his wife and three children. If I am told, that the farmers, that the occupiers of houses and land, are so poor that they cannot do more for their wretched work-people and neighbours; then I answer and say, What a selfish, what a dastardly wretch is he, who is not ready to do all he can to change this disgraceful, this horrible state of things!

72. But, at any rate, is the salary of the "ASSISTANT OVERSEER" necessary? Cannot that be dispensed with? Must he have as much as all the widows, or all the old men? And his salary, together with the charge for printing and other his various expenses, will come to a great deal more than go to all the widows and old men too! Why not, then, do without him, and double the allowance to these poor old women, or poor old men, who have spent their strength in raising crops in the parish? I went to see with my own eyes some of the "parish houses," as they are called; that is to say, the places where the select vestry put the poor people into to live. Never did my eyes before alight on such scenes of wretchedness! There was one place, about 18 feet long and 10 wide, in which I found the wife of ISAAC HOLDEN, which, when all were at home, had to contain nineteen persons; and into which, I solemnly declare, I would not put 19 pigs, even if well-bedded with straw. Another place was shown me by JOB WALDRON'S daughter; another by Thomas Carey's wife. The bare ground, and that in holes too, was the floor in both these places. The windows broken, and the holes stuffed with rags, or covered with rotten bits of board. Great openings in the walls, parts of which were fallen down, and the places stopped with hurdles and straw. The thatch rotten, the chimneys leaning, the doors but bits of doors, the sleeping holes shocking both to sight and smell; and, indeed, every-thing seeming to say: "These are the abodes of wretchedness, which, to be believed possible, must be seen and felt: these are the abodes of the descendants of those amongst whom beef, pork, mutton and veal were the food of the poorer sort; to this are come, at last, the descendants of those common people of England, who, FORTESCUE tells us, were clothed throughout in good woollens, whose bedding, and other furniture in their houses, were of wool, and that in great store, and who were well provided with all sorts of household goods, every one having all things that conduce to make life easy and happy!"

73. I have now, my friends of Preston, amply proved, that what I have stated relative to the present state of, and allowances to, the labourers is TRUE. And now we are to do all we can to remove the evil; for, removed the evil must be, or England must be sunk for ages; and, never will the evil be removed, until its causes, remote as well as near, be all clearly ascertained. With my best wishes for the health and happiness of you all,

I remain, Your faithful friend, and most obedient servant, WM. COBBETT.

THE END.

Printed in Great Britain
by Amazon